Experimental Design and Scientific Data Analysis

AF833925

Experimental Design and Scientific Data Analysis provides accessible information about the research process from start to finish as applied to conservation to enable readers to successfully plan and carry out research and share findings effectively with others.

The book is divided into two parts, starting with experimental design in Part I and then considering data analysis in Part II. Part I takes the reader through the process of designing and planning experiments, from the initial design of aims and objectives to ethical testing, and includes a range of examples to demonstrate how these ideas apply to a range of specialisms and materials. Part II focuses on what to do with data once it has been collected, including choosing the right statistical test, different methods of presenting information, and communicating your findings clearly to your chosen audience. Questions at the end of each chapter are designed to help readers check and consolidate their knowledge of the different topics.

Experimental Design and Scientific Data Analysis is suitable for students, emerging professionals, and experienced conservators at different stages of their careers, and is particularly suitable for those without a scientific background.

Lynda K. Skipper is a senior lecturer in conservation at the University of Lincoln, UK, focusing on science, research, and preventive conservation. She has a PGCE in higher education teaching and is a Fellow of the Higher Education Academy. She researches historic interiors and stone biodeterioration.

Philip J. A. Skipper is a research associate and associate lecturer at the University of Lincoln. He also works as a freelance consultant, currently contracted with Historic England to provide advice and carry out research projects on biological growth, biodeterioration of heritage materials, surface treatments, and general heritage science.

Science for Conservators
Series editor: Joyce H. Townsend

The *Science for Conservators* series is now being revised and fully updated, and it will be extended in scope. The original series has provided key basic texts for conservators throughout the world since its publication in the late 1980s. Scientific concepts are basic to the conservation of artefacts of every type, and these introductory volumes provide an essential theoretical background for conservators who have entered the field without scientific training. It is designed for pre-programme, student and personal study, and also serves as a checklist of scientific terms for those working in English as a second or third language.

An Introduction to Materials and Chemistry, third edition
Joyce H. Townsend

Colour, Colour Measurement and Colour Change
David Saunders

Experimental Design and Scientific Data Analysis
Lynda K. Skipper and Philip J. A. Skipper

For more information about this series, please visit: www.routledge.com/Science-for-Conservators/book-series/SFC

Experimental Design and Scientific Data Analysis

Lynda K. Skipper and Philip J. A. Skipper

Routledge
Taylor & Francis Group
LONDON AND NEW YORK

Designed cover image: by Lynda and Philip Skipper.

First published 2025
by Routledge
4 Park Square, Milton Park, Abingdon, Oxon OX14 4RN

and by Routledge
605 Third Avenue, New York, NY 10158

Routledge is an imprint of the Taylor & Francis Group, an informa business

© 2025 Lynda K. Skipper and Philip J. A. Skipper

The right of Lynda K. Skipper and Philip J. A. Skipper to be identified as authors of this work has been asserted in accordance with sections 77 and 78 of the Copyright, Designs and Patents Act 1988.

All rights reserved. No part of this book may be reprinted or reproduced or utilised in any form or by any electronic, mechanical, or other means, now known or hereafter invented, including photocopying and recording, or in any information storage or retrieval system, without permission in writing from the publishers.

Trademark notice: Product or corporate names may be trademarks or registered trademarks, and are used only for identification and explanation without intent to infringe.

British Library Cataloguing-in-Publication Data
A catalogue record for this book is available from the British Library

ISBN: 978-1-032-50164-2 (hbk)
ISBN: 978-1-032-50163-5 (pbk)
ISBN: 978-1-003-39713-7 (ebk)

DOI: 10.4324/9781003397137

Typeset in Times New Roman
by Apex CoVantage, LLC

This book is dedicated to Muriel and John Fuller, whose encouragement and support has been invaluable since I was a child. Thank you for everything.

– Lynda

To Lynda, without whom I wouldn't have written part of this book; also to Elliot, Ben, Ofsi and Socks for giving us both the time to write.

– Philip

Contents

Acknowledgements

Many thanks to our students past and present, who inspired us to write this book. We are grateful to the series editor, Dr Joyce Townsend, for her comments on our draft and to our reviewers who provided input on our initial proposal.

Figures

Tables

Foreword to the *Science for Conservators* series

The series was first published in 1982 as a series of three books aimed at conservation students and conservators who had come into the field without any scientific training beyond their high school years. It was intended to develop into some six or seven topics covering a broader spectrum of knowledge than the first three, which concentrated greatly on chemistry. The series of three was reprinted in 1987 and again in 1992. It has been in steady demand ever since, despite the text being unchanged since the early 1980s. Some conservation training programmes in a range of countries have recommended the series for pre-programme study, while others have included it in their first-year reading list for the students from a humanities background or for all their students.

Having no training in the natural sciences after the early teenage years, and sometimes even fewer encounters with the biological sciences, had been the background of most conservators in all countries for much of the 20th century. It still is for some pre-programme conservation students from a few countries, for those who undertake apprentice-style practical training straight after high school, and for others who enter the field of cultural heritage conservation through post-graduate training that was preceded by humanities-based or social sciences–based graduate study. The conservation of artefacts is a truly inter-disciplinary activity which would best be begun with the knowledge of a lifetime! Most conservation training programmes today, including those for heritage scientists, aim to provide pathways into thought processes and knowledge bases from the natural sciences, the humanities generally, and to a lesser extent the social sciences, engineering, the biological sciences, and philosophy and ethics as well, through targeted teaching in all these areas.

The fully revised and updated *Science for Conservators* series is building on this concept. It is intended to serve as a basic primer and an aid to the assimilation of concepts that will become embedded in the thought processes involved in decision-making by conservation professionals. The series is targeted at several groups: those about to enter a conservation programme without personally having a scientific background; conservation professionals in training and with a similar background; those who are establishing themselves in their challenging and rewarding field; professional conservators who may want to check that they are using a scientific term to mean the same thing a scientist would assume it to mean rather than its meaning(s) in everyday conversation; and those who have studied the natural sciences or even the conservation of artefacts in another language than English but are now working, writing reports, and publishing in English.

When reading each book, allow yourself to become completely familiar with a section and confident about its contents before moving on to the next. Do not read large portions at any one sitting. Although the series is an elementary one, you will need to take plenty of time in working all the way through it. You should not feel disheartened if your progress at times seems slow. If you do have particular difficulty with a section, ask a conservation scientist, a heritage scientist, or another conservator about it. It is not worth struggling on your own; even a scientist with no knowledge of conservation can help, although one associated with a conservation programme or a museum would be even better. Very often, the problem seems surprisingly simple to clear away if you can go through it with somebody else.

Worked examples and exercises have been included where they will be useful. Check your answers at the end of the book. Occasionally, some simple demonstrations are suggested to illustrate or clarify the written text. At relevant points, you will also find reference tables. When scientific terms appear or are defined for the first time, they are **emphasised thus** for easy reference. A full index of such terms is included at the end of each book.

Knowledge in the conservation profession is always growing. New journals are established, and the longer-running journals can now be accessed online right back to their first volume. Conferences and events take place every year and are now preserved as recordings of the presentations and discussion sessions as well as in published conference proceedings. There are hundreds of books now written by conservation professionals for their fellow professionals, with much less need for conservators to rely on textbooks written for pure scientists. I am planning that this series will be expanded beyond the three books originally published, to cover a broad spectrum of topics relevant to the conservation of artefacts – just like the original concept of the (short) 1982 series. Further books now include *Colour, Colour Measurement and Colour Change*, *Experimental Design and Scientific Data Analysis* (this book), and *Preventive Conservation in Practice.*

Knowledge of health and safety has changed and developed a great deal since the 1980s, of course. The conservation profession has also evaluated the effects of past treatments on the preservation, appearance, and, indeed, survival of objects, and was beginning to do this, and publish such evaluations, in the 1980s. Many treatments that were common before and in the 1980s, and some chemicals and reagents that were then available on the bench in most conservation workspaces and every undergraduate chemistry laboratory, now appear hazardous to objects, dangerous to the user, and inappropriate to use in the third decade of the 21st century. Yet they played a part in the long history of a great many artefacts: it is necessary to know what might have been done to an object in the past in order to treat and sometimes stabilise it now. Also, treatment records used the terminology of their era. I have taken the decision not to exclude from the updated series the materials that are no longer used by responsible conservation professionals. They were used at the time for their unique properties and perceived efficacy, and they may have affected the properties of an object and its likelihood (or not) of surviving for another few centuries. The reasons they fell out of regular use will, however, be mentioned.

Some materials derived from petrochemicals are going to be much less used in the future, as greener alternatives are developed and come into wider use. Green alternatives made locally will become cheaper than imported chemicals or those made from petrochemicals. Using greener materials and cutting down on both waste and transport are very active topics of discussion in the field of conservation at present, and sustainability aligns with the profession's core values too. In this new text, the materials falling out of use for such valid reasons as health and safety considerations and the green movement will be shown in grey font to emphasise that they have or will recede from use.

Introduction to this book

As a student myself, and then moving on to teach conservation, it was clear to me that students were starting their conservation journey with two quite different outlooks on science. Some people love the idea of science-based lab work and are keen to get started testing materials and carrying out experiments. Others have perhaps had a difficult time with science at school, and it's not unusual for me to hear the phrase 'I can't do science'. For people who fall into the latter category, this book will help you to feel more confident about scientific aspects of conservation and how to present work that you've done. For those who do feel more confident, I hope this book will give you new ideas and expand your knowledge so that by the end of the book, everyone – no matter what your starting point – will not only feel able to 'do science' but to do science *well.*

You may be wondering why that's important; what does a conservator need science for? In my opinion, science is integral to everything that a conservator does and underpins all of our work. Some conservators use more science than others, but having a scientific understanding is essential. Sometimes, you may not even realise that you're using scientific techniques. For example, when you are deciding what type of solvent to use to remove some surface dirt, you might carry out solvent testing to try different concentrations of a selection of solvents. Perhaps you are trying to dye a textile and need to decide on the best dye ratio to get a good colour match for netting over some shredding silk. When you look at a metal, you might try to identify it by colour and weight and corrosion colour – but what if you need to know the proportions of copper and tin in a bronze statue or to work out whether a pewter is an older, lead-based material? For all these examples, you will need to take a scientific approach.

The scientific understanding of objects and built heritage feeds into decision-making for that item or collection. It can contribute to the story of that object or building, through a clearer understanding of materials and manufacturing. It can enhance significance and give a deeper understanding of the object's research **value** as part of **values-based decision-making**. Using science to understand how to appropriately conserve the materials of the object may enhance its aesthetic values or educational values. Knowledge of the object or heritage site obtained through **scientific methods** may enrich grant bids, either externally or when requesting funding from within your organisation. It's not just individual objects

DOI: 10.4324/9781003397137-1

either; effective presentation of data about preventive conservation, for example, can improve reporting of the work that conservators do and help make information more accessible to non-conservators – whether that's a museum manager, a facilities manager, or as part of public outreach.

Ultimately, science is a creative process, and the experimental design and process of carrying out testing is a craft skill. When conservators start to consider scientific technique as an extension of art-based practices, it becomes clear that science simply applies existing conservation skills of problem-solving, creativity, **observation** and pattern recognition, planning projects and of course practical skill in a slightly different way. When conservators investigate objects, they are often aiming to solve problems, driven by their curiosity to find out more; the same curiosity which motivates those in scientific roles. The skills and understanding of materials or objects that conservators have put them in a strong position to make a valuable contribution to scientific knowledge and understanding in their chosen specialism.

In summary, I would argue that science is always an integral part of a conservator's job role. In fact, it's something that a conservator will be doing most days without even perhaps thinking about it as science. However, there are also times when you will need to plan and construct more in-depth testing or analysis and to present that in a suitable format. The different chapters in this book are designed to help you with all stages of this process. The chapters can be read as stand-alone sections for help with a specific aspect, or you can work through the book sequentially to further develop your overall understanding of experimental design and data analysis and how to apply it.

– Lynda

Book structure and how to use it

This book is structured in two parts; each part is split into three chapters. The first part considers how to design an experiment in a way that is repeatable and valid and allows you to reach meaningful conclusions about the things that you are trying to test.

Our first chapter discusses **Why?: writing a research question**. This chapter starts by exploring in more detail the reasons why conservators might need to design experiments, giving some examples. This moves into discussing the initial planning stage of experimental design, thinking about which question(s) to ask as part of the research, creating a hypothesis, and then designing appropriate aims and objectives that work to answer that brief. It's also important to ask whether your testing is necessary and what the purpose is for your work. Understanding your own 'why' of research will help direct your planning and experimental design to make sure everything is relevant.

Chapter 2 leads into planning appropriate techniques as part of **How?: experimental design methodology.** A vast range of scientific techniques are available, and you need to consider which one(s) to use before any testing can begin. This chapter explores different types of data, both qualitative and quantitative, and the benefits of a mixed-methodology approach. This chapter does not aim to be a comprehensive guide to all analytical techniques that are currently available. Instead, we give an overview of the general principles of the different techniques and the types of data, with some examples of how techniques might be used to generate this type of data including visual information, such as microscopy, simple spot testing, and more complex analytical techniques including elemental composition and chemical structures analysis.

In Chapter 3, we move on to look at **What?: ethical sampling and scientific rigour.** This chapter considers some current conservation issues, with a section on ethical sampling practice and decision-making, exploring the risks and benefits of directly testing on historic objects versus the use of surrogate materials. The idea of scientific rigour and ensuring data is valid is discussed here, including the concept of controls and (where possible) repeated testing to reduce the risk of error. The principle of 'humble science' is an important part of reflective practice, and we use this concept to consider how people need to be open to making mistakes and the levels of potential uncertainty in the interpretation of results.

Having used the first part of this book to design your research, we can assume you may now have data or are in the process of collecting it; but what are you going to do with it? Ideally, this is something that you consider as part of the experimental design stage, before it is then interpreted to produce the output of your research. This is discussed within **Part II: Scientific reporting and data handling.** Chapter 4 of the book gives an **Introduction to statistics for conservators**. It introduces the concept of statistical methodologies through analysing data and choosing the right statistical method based on the type of data that's been collected and how to use terminology to describe it correctly. This chapter aims to give conservators the tools to better understand statistics used in papers they are reading and to help them to choose appropriate techniques for their own studies. Conservators are also encouraged to know their own limitations and when it may be more appropriate to collaborate with data scientists to enhance their work.

Chapter 5 presents ideas of **How to display results.** Many people struggle with deciding on the right way to show data in an accessible format for their audience. This chapter discusses options around presenting data, including the main types of graphs and when to use tables. This chapter is not intended to give people training on how to produce these using a computer programme; rather, it is intended to help people make decisions about what format to use. Results may also be presented as photographs and/or images, and recording and labelling of images are discussed here too, as well as the production of meaningful image captions. Displaying results forms part of documenting the research, and this is expanded further in the final chapter of the book.

In Chapter 6, **Writing reports in scientific style** will develop your ability in writing reports for a range of different purposes e.g. client, journal or conference paper, university assessment, or funding application for a grant. Within these different purposes, there are some standardised approaches which will be explored in detail, breaking a report down into what is included in the different sections such as methods, results, and conclusion. This will relate to Chapter 1 in terms of feeding back into an initial aim for the research – while also acknowledging that sometimes, the results may show something unexpected!

Finally, you will find a **glossary**, which you can use to help your understanding of scientific terms and concepts as needed. Words in the glossary are highlighted in blue in the text.

Within each chapter, you will see worked examples and exercises to help you check your understanding of the different topics. There are some cues to sources of further reading where you can explore topics of interest in more depth. You can also make use of our supplemental material online, which contains extra learning materials that are intended for individuals or for use as exercises within classes as part of training courses. These will help you consolidate your knowledge and review what you have learned.

As a final thought to close this introduction – try to enjoy the science, and don't be overly critical of what you do! Science-based research may seem daunting at first, but it's a skill everyone can learn through practice. Think back to the first time you tried playing a musical instrument or the first time you baked a cake. It won't be perfect the first time, and you will make mistakes, but those mistakes all offer you chances to learn and improve. Science works in the same way, and even the best scientists make mistakes or have to deal with things that don't go as planned.

Sending virtual encouragement to everyone,

Lynda and Philip

Part I

Experimental design

1 Why?

Scientific method and writing a research question

Introduction to conservation research

Research is defined as a process that leads to the discovery of new knowledge. Research in conservation can cover a wide range of different topics and methods that are used by conservators and heritage scientists to find out more about objects, materials, historic sites, and the environment around them. Conservation research is key to enhancing conservation knowledge and to developing the profession. Scientific analysis of cultural heritage can aid understanding of **values** and significance of objects and broader cultural and historic perceptions, such as analysis of glass to determine place of origin, and therefore increase understanding of historic trade routes. Scientific experimentation may also help conservators deal with a range of emerging challenges, such as movement of pest species due to changing climate or investigation of new conservation materials and techniques that perhaps work more effectively, are suitable for newer materials such as plastics, are suitable for use outdoors on sites or in hot/cold climates, or have a lower environmental impact.

This book sets out the stages of a research process known as the **scientific method**. The scientific method can be broken down into a series of steps (Figure 1.1). An initial **observation** acts as inspiration for the area of research, and further refining of the topic leads to the production of the **hypothesis**, **aims**, and **objectives**. These are then tested through experimentation to produce data, which is analysed to draw a **conclusion**. These conclusions may then feed into further observations and future research topics.

Although this book focusses on the scientific and experimental aspect of research, some of the topics we discuss will also be useful for planning and communicating a wide range of different forms of research. Research in conservation is certainly not limited to scientific experimentation. It covers a diverse variety of methods including questionnaires and focus groups, archival, documentary, and literature-based research, object-based research, and investigation of historic production technologies. Conservation research is a valuable methodological approach because of the direct engagement with the physicality of objects and buildings, alongside the other methodologies. Clark (2001) acknowledges this complexity and defines the term **CoBRA (conservation-based research and analysis)** as 'the research, analysis, survey and investigation necessary to

DOI: 10.4324/9781003397137-3

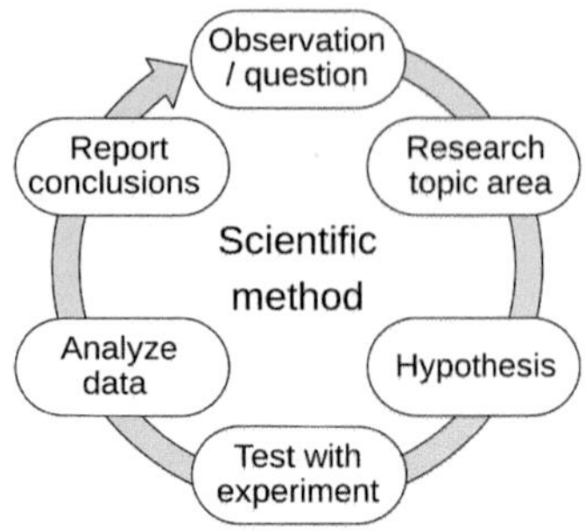

Figure 1.1 Stages of the scientific method, viewed as a cyclical process. Image by Efbrazil, CC BY-SA 4.0 <https://creativecommons.org/licenses/by-sa/4.0>, via Wikimedia Commons.

understand the significance of a building and its landscape, and thus inform decisions about repair, alteration, use and management'. Although Clark's definition is focussed on built heritage, we would argue that the definition should be broadened to include research pertaining to objects and materials and how that research aids understanding of significance and informs decision-making in conservation, irrespective of subject specialism or topic.

It's important to remember that the *value* of the research that's being carried out is dependent on the quality and rigour of the experimentation or other form(s) of data collection (concepts of scientific rigour and validity will be explored in more detail in Chapter 3). It's not just about doing research, it's about how to do *good* research. In order to design a successful research project, it's essential to have a plan. Part I of this book aims to help you with that planning stage through setting out a framework for developing a successful research project, starting with defining the reasoning and justification for the research itself. Being prepared for the research before you start and having a very clear idea of where you want that research to go increases the chance of the research producing meaningful and valuable results. However, it's also important to acknowledge that it's not possible to plan for every eventuality, and the process of doing the research itself and then analysing results is also part of the learning process. Sometimes, as part of the learning process, things don't go as planned, and adjustments to the topic, experimental design, or timescale might be needed to get back on track. We don't think anyone involved in scientific research could claim that their experiments always work the first time, every time, or that their first statement of a research question was the best and most complete question to ask in the context of the treatment.

All research is expected to meet certain standards. Research in a peer-reviewed journal will only be published if the authors show that they have met specific criteria, which means it can provide a useful contribution to understanding of that topic. The areas listed in box 1.1 all need to be considered at the planning stage to produce a valid outcome that contributes to understanding of that topic.

Box 1.1 Expectations for research standards in conservation. Modified from text produced by Leah Warriner-Wood, personal communication, reproduced with permission.

- *Rigorous and systematic: You need to have a clearly defined purpose; this will include a clear research question and repeatable plan for your research methodology.*
- *Valid and current: If the topic is one which has been covered in the past, then your research project should build on prior work rather than just repeat it. Your research should add to knowledge of a field or topic. (A review paper, which is defined as a summary of research in the field at present, can still add new knowledge through bringing together viewpoints to create a core reference guide to a topic.)*
- *Reliable and open to public scrutiny: your research should use valid sources to back up its findings/conclusions and be transparent in its use of other people's research.*
- *Conform to accepted methodologies: 'because I think so' or 'I asked my Mum/Dad/partner and they said . . .' are not valid research methodologies and won't stand up to scrutiny (unless your Mum/Dad/partner just happens to be a published expert in your topic, of course)!*
- *Conform to accepted formatting standards: these can (and will) vary from institution to institution/publication to publication, and you should always consult the guidelines provided before you begin.*

1.1 Why? how? and what? in research

Start with Why (Sinek, 2011) considers leadership and how leaders can inspire people to act. He discusses the 'Golden Circle', a concept inspired by the golden ratio, which is a mathematical formula for proportion. When applied to leadership, he argues that the 'why?' circle is central, followed by 'how?' and 'what?' (Figure 1.2). Sinek advocates that having a different perspective to help us organise human behaviour helps us understand why some leaders (and brands) are particularly successful. When we receive communications from companies, they often focus on 'what?', which is relatively easy to articulate. He argues that starting from the centre of the circle and building outwards is a more effective way of communicating, through building understanding and loyalty to an idea, or set of values. The Golden Circle principle has much wider applications than marketing or leadership, though. We wrote the structure for this book centred around this principle and the importance of starting your research with 'why?' Through having a clear justification and basis for your research, you have a foundation from which you then build your how? and what? (Chapters 2 and 3). In this book, conservation science and experimental design follows the Golden Circle. We will start with why?.

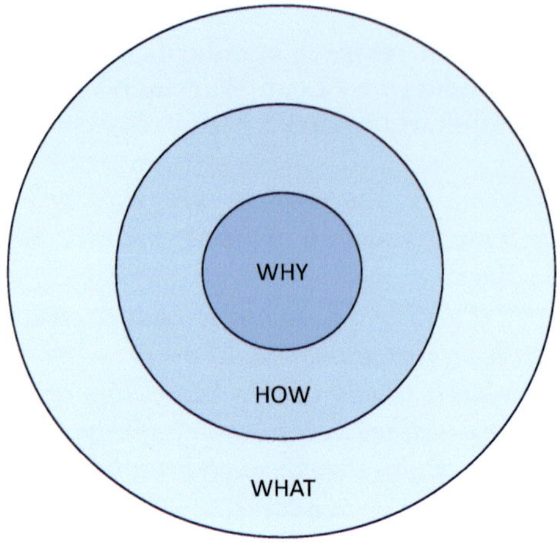

Figure 1.2 Golden circle concept with why in the centre, surrounded by a circle of how, and then what. Image by Lynda K. Skipper.

1.2 'Why?' begins with a research question

Research question is a term used in academia to refer to an overview of the reasoning behind the topic that someone is researching. It articulates a particular problem, and the data gathered through the research is designed to address that problem. In other words, research data in whatever format is presented to answer the research question posed at the beginning and helps the researcher form a conclusion. Using the phrase 'research question' does shorten that fairly lengthy explanation, but because of the word 'question' within it, it sometimes leads people to think that all research questions need a question mark on the end. There are certainly times when ending a research question with a question mark is appropriate, and we'll explore some examples in this section, but equally, there are times when it's not necessary to structure a research question in a question format – a statement might be more effective. We will use the term 'research question' through this book because it's in common use, and we will start by delving deeper into its purpose and then design.

1.2.1 Why is a research question important?

A research question gives the background justification for your research. It sets out the issue succinctly and justifies the reason for doing the research. This might be a particular problem or challenge that the researcher has encountered in their work, or it might be commissioned/set by an external body. Without a strong research question, funding bids can be too vague, or data gathered may not fully address what the researcher hoped to find out. That research question is the starting point for building an **argument** when the results are being written up; this will be discussed in more detail in Chapters 5 and 6 when we discuss report writing and presentation of data. Research questions are unlikely to be right the first time; they

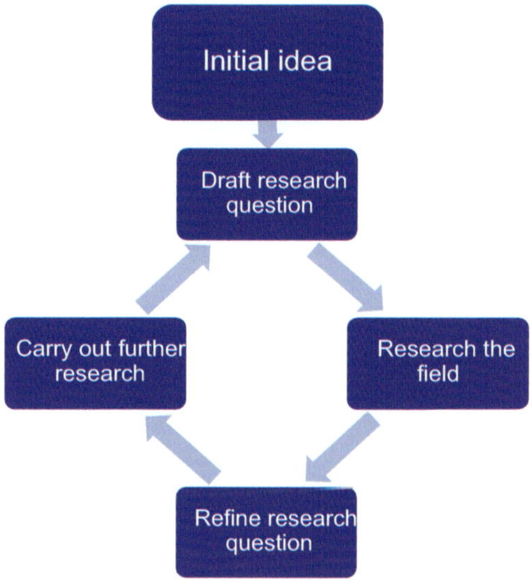

Figure 1.3 Stages of drafting and refining a research question. Image by Lynda K. Skipper.

always go through a process of development, and like the scientific method shown in Figure 1.1, this development should be viewed as a feedback loop (Figure 1.3).

After initially drafting and refining a research question, trying to write aims and objectives (section 1.3) will explore whether that question is precise enough or if it needs further refining. With adjustments to the research question, the aims and objectives can then be further improved. Literature-based research feeds into the process through understanding what work has been done in the past and how your own research might add new knowledge. As research progresses, your plans may need to be revisited and refined based perhaps on findings from a **pilot study**, from unexpected results, or through new opportunities to access different materials or equipment to enhance the study. Therefore, while having a defined starting point for the research is very important, it should not be viewed as something that is always static, but instead, research is an iterative process that may need to adapt and change with new knowledge or changing circumstances.

As part of the design phase, it is useful to consider the underlying research paradigm (box 1.2), which will influence how you approach the study.

Box 1.2 Research paradigms

The way in which you approach a research project is known as a ***research paradigm****.*

Positivism tends to be most strongly associated with a scientific approach to collecting data. It is based on collecting information which can be measured and assessed in an objective way, without bias. This paradigm implies that we cannot be confident of a fact if it is based on something that cannot be measured.

Interpretivism assumes reality is a social construct and that information is subjective. This paradigm covers data which cannot be measured, for example different viewpoints of a historical event, and knowledge is perhaps gained through discussions, interviews, or close reading of historical documents.

Pragmatism tends to take a more mixed viewpoint and takes the broadest possible viewpoint when considering a problem. With a pragmatist paradigm, the research may not lead to certainty because meaning is a construct of human experience, and what is known now may change over time as new information becomes available. This is arguably the more typical paradigm for conservation-based research which involves a mixture of data types.

1.2.2 How to find your research question

Some readers may be approaching this chapter with a research topic in mind. Perhaps they have come across something in their workplace that intrigues them, and they want to explore it in more detail. Some conservation students who have been asked to write a dissertation on a topic that interests them may have a much broader starting point and a less clear idea of what their research question actually is.

As a starting point, we have always found spider diagrams or mind maps to be very helpful. They work well for creating a visual representation for a range of different ideas that expand from one central theme and can be produced on paper or on a computer. Let's assume we have a conservator who wishes to research something about gilding but doesn't have a clearer path than that at this stage. Figure 1.4 shows an example mind map that might be used as a starting point for that conservator to develop and refine a research idea. There are four broad areas coming from the central theme. These topics are quite broad, then more lines are added to extend the ideas, narrowing them down and becoming a little more specific. This is not yet a research question at this stage, but this approach helps to formulate some topics which could be explored further.

Following the example of gilding a little further to illustrate how someone might refine a research question, let's consider one aspect: that of bole. For people unfamiliar with gilding techniques, bole is a clay used in water gilding to coat the surface after applying gesso. Its colour affected the colour of the gilded surface. It's traditionally used on frames and a range of other gilded surfaces.

There are many different research questions that could be explored around bole, which could form additional branches from the mind map in Figure 1.4. We will assume that after extending our mind map, the researcher wishes to focus on elemental analysis of bole to research its composition. The researcher will then

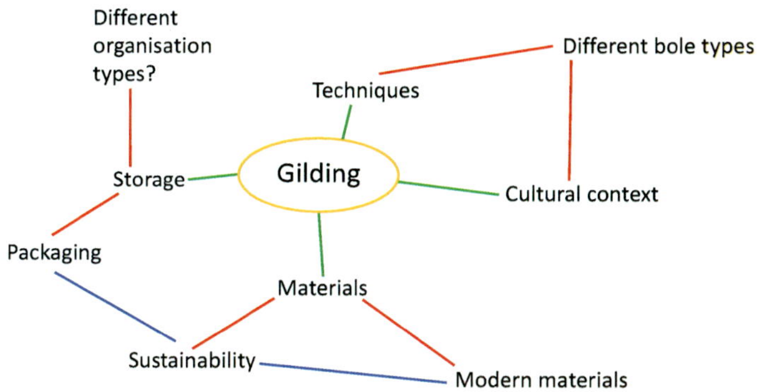

Figure 1.4 Example of a mind map based on the central theme of gilding. Four concepts lead directly from this (green lines), which then inspire other connections (red lines). Interconnections between some concepts are shown as blue lines. Image by Lynda K. Skipper.

need to consider why they are interested in this – to better articulate the 'why?' aspect of the research. Why is the research justified, and what aspect of knowledge might it enhance? There is a range of potential reasons for this; perhaps to improve decision-making in repair of gilded frames, through a better understanding of materials. Perhaps the researcher wishes to explore the origins of bole, understanding whether bole was sourced locally to the place of production or from different countries. The reasoning behind the research may mean that the experimentation itself is conducted in a different way, through choice of the materials for testing, for example. In the first example, comparing a range of historic bole types to modern bole might offer an interesting comparison. With the latter example, the researcher may wish to select frames with a range of known geographical origins in a similar time period or from one place over a range of different eras to see how bole types changed over time.

As a thought experiment, why not pause for a moment and consider whether you might be able to suggest any other justifications for carrying out this research topic?

The next stage of the process is to refine the research question further, making it precise and detailed enough that it can clearly inform the 'how and what' aspect of research part of the planning. This is not a process that can happen in isolation. To turn these initial thoughts and ideas into a research question that will clearly lead to new knowledge, you will need to research what knowledge is already published so that you avoid repeating the work of others.

1.2.3 Using literature to refine and shape the research question

At the start of this chapter, research was defined as a process leading to the creation of new knowledge. In the planning stage, it's therefore important to know what work already exists. Continuing with our example from the previous section, if another researcher has already analysed bole from Italian frames from 1655–1695 and published the results, would extending the timeline by a few more years really be useful? Alternatively, perhaps the data already exists but has not been fully utilised. For example, analysis of bole composition for the frames in a collection may be appended to the collections database for each individual frame but has not yet been collated together and analysed systematically with that research question in mind. All this background information will help to shape the approach to the overall project – or potentially lead to rethinking of the topic. While needing to rethink the topic is frustrating and can be disheartening, it is better to reach that decision at the planning stage than to realise that there are significant issues with the topic when the project is already underway.

Any research project should always incorporate a **literature review**. When developing the project, reviewing literature that has been published in the field is used to demonstrate what's already been done and helps you to find areas in which there are gaps in understanding. Being able to search outside your immediate field of work is also a valuable skill; techniques may be used in one area which could, after testing and research, be applied to another field of conservation. Many published articles will mention further work – areas which they have not been able to cover within their project – and this can be a potential source of inspiration for people who are struggling to find a topic of interest.

At this early stage of development, information can be read quite strategically. In a paper, we would always start with reading the **abstract**. If it's not relevant to our current project plans, we may file it for later reading – but then swiftly move on to the next one. If the abstract suggests it would be relevant, then we would read more of the paper. The introduction section of papers will often point towards other relevant references covering related topics in the field, so one relevant paper can be a good source of further information. The methods section of a paper can help with developing the methodology of your own work (see Chapter 2). The results and conclusion are relevant for showing exactly what was discovered and how information from that research might be relevant to your own areas of interest.

How do people find these sources of information? Different organisations will have different levels of access to publications, and private (or retired) conservators may have limited access to information that is behind a paywall. Conservation books can be costly and are not always available in a range of different languages. This can be a challenge and may introduce inequalities of access in the field. The recent move towards publishing all articles as open access is a good starting point to help address this issue. However, open-access journals tend to charge the authors a significant fee for publishing, which can be a barrier to dissemination of work. This is something that researchers should be aware of, as it may present challenges in information gathering to support their research. It is worth noting that many university libraries allow associate membership for a small annual fee, which can be a way to access books and, in some cases, journal articles; similarly, online communities such as ResearchGate and Academia can allow you to request copies of journal articles from the authors.

If we assume that the researcher has some access to information to carry out this search, **keywords** are used to try and find relevant information and sometimes to filter out or exclude irrelevant aspects. Conservation is particularly difficult as a search term, given the overlap with environmental conservation. A search for wood conservation is likely to bring up a range of sources that have more to do with managing forests than with furniture! As such, searching for 'preservation of heritage furniture' is more likely to provide relevant results than 'wood conservation' as a search term. The use of AND to include terms, NOT to exclude terms, and OR to search for alternatives can help with filtering if your search engine offers that functionality. Use the exercise in box 1.3 to explore the concept of keywords further.

Box 1.3

Imagine that you are trying to search for the analysis on the use of gels for removing staining in furniture conservation on unvarnished wood. Take 2 minutes to note down your thoughts. Using your preferred search engine, try out the key words, and decide which ones (either singly or in combination) are more effective for producing valid results.

There are no answers provided for this exercise because there are no right or wrong answers; the key learning from this is reflecting on what works and why and testing different ways to refine search terms to find ones that are suitable for the topic which is being investigated.

Literature reviews provide a basis for all research, whether it's based on scientific experimental or other research types. A literature search will help ensure that your initial research question is valid, and from there, the researcher can start to develop the project further, through building a set of aims and objectives. Your literature review will also form part of the report, which we discuss further in Chapter 6.

1.2.4 Refining a research question

Integrating knowledge from literature and background research is a vital part of the process of refining a research question. A greater understanding of the topic and what's been covered already helps to identify the gaps, where you can add new knowledge. As shown in Figure 1.5, by moving through a series of developmental steps, we will reach the final stages of refining a research question, which involves turning the overall concept of the research into a defined research problem (arrow 4). By revisiting our gilding example once more, a way of phrasing this as a statement rather than a question could be: *Determining the geographical origins of bole used on Italian water gilded frames from 1650 to 1700 via elemental analysis.*

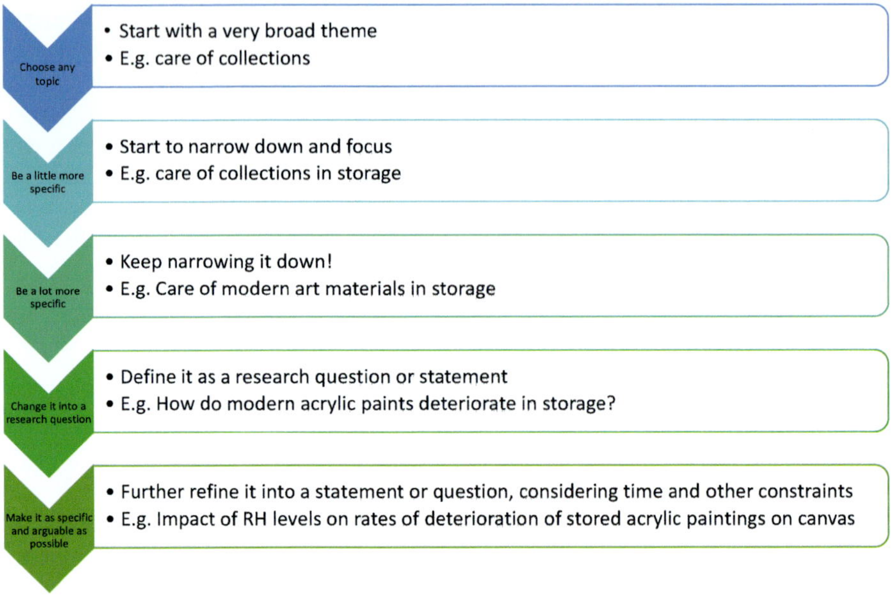

Figure 1.5 Stages of refining a research question. Image by Lynda K. Skipper.

However, this can also be changed into a question format, depending on the focus of the research, and sometimes the level of certainty or confidence in the potential for results. For example, *Can elemental analysis of bole composition on Italian frames from 1650–1700 aid understanding of the material's origins?*

These steps all help to further focus the research question, but to fully shape this into a project that is practical instead of aspirational, the researcher needs to consider any constraints around the research (arrow 5, Figure 1.5). A research question needs to take into account available resources or constraints, which may include (but are not limited to) funding, person-hours, or access to materials for study. Those constraints will be individual to the researcher's circumstances and the type of project taking place. An undergraduate dissertation project, for example, would be expected to be less complex and smaller in terms of what can be achieved compared to a three-year PhD project. The idea of constraints and how they may impact decision-making is discussed further in Chapter 2, section 2.4.

When the researcher has a research question which sounds suitable, then it is time to start writing aims and objectives.

1.3 Writing aims and objectives

Aims and **objectives** are an essential step in firming up an initial idea for a research topic. They help give some more clarity to the research plan. With large projects in particular, objectives help to break down the project into defined smaller steps, which can make it feel like a more manageable prospect. Reviewing your aims and objectives will help with tracking progress and making sure the research you're doing is following its original purpose – they will help you focus on the topic at hand. We know from experience how easy it is to get sidetracked into a side project because you happen to have noticed something interesting; it may be broadly related to the research, but if it doesn't help the core project, then aims and objectives will help with deciding whether it's relevant to pursue it now or whether it's something to put to one side for another time. They will also feed into your overall project report or other output, which is discussed in Chapter 6.

1.3.1 What are aims and objectives?

Aims are designed to be a broad statement – or statements – that explore what you intend to show or test during your research. They look at the bigger picture and the overall idea of where you think the research will go. They will arise from your overall research question, which was discussed in section 1.2. Aims should be written carefully so as not to pre-judge the outcome. If you set out to *prove* that a particular new conservation technique is better than existing methods, it can risk introducing bias because you are looking for results that fit your aim, and you might dismiss results that don't support what you are hoping for. **Bias** is something that you can read more about in Chapter 3.

Objectives are the steps which are needed to complete that aim, and there will often be 2 or 3 objectives that feed into one aim. These need to be SMART. If you haven't come across this acronym before, **SMART** stands for **specific, measurable, achievable, relevant, and time-bound**. This concept was first introduced in relation to project planning in the 1980s, but it has much wider applications, including planning research projects (research projects can also be successfully managed through considering project management frameworks).

SMART objectives are precisely worded, with a clear end point, so that the researcher can look at them and understand that they have reached an end point for that part of the research. As an example, an objective to 'measure rainfall' would be far too vague. How long is the researcher measuring rainfall for? And where? The whole world or just a specific heritage site? To make this into a SMART objective, the researcher would need to consider narrowing down the focus and being more specific. A revised objective could be 'Measure levels of daily rainfall over the course of 12 months at Lincoln Cathedral'.

In order to show how the planned research meets SMART goals, it's useful to dissect the individual aspects of each objective as part of the development process (Table 1.1). When we start to break down our example objective further, it reveals some questions that would need to be considered by the project team if this were an objective for a real research project. These would not necessarily all be answered within the objective itself but would need to be addressed in the research planning stages.

Table 1.1 Dissecting the information in the objective to measure levels of daily rainfall over the course of 12 months at Lincoln Cathedral using a SMART approach.

SMART approaches	*How the objective meets that SMART requirement*
Specific	It is fairly clear which site is being monitored and what is being measured, though there are multiple places called Lincoln across the world, and defining UK as the place for monitoring within the objective would avoid any possible confusion.
Measurable	The rainfall is being measured over one year, so it is clear when the objective will be reached, and there is a defined 'thing' which is being measured. At this point, it doesn't clarify how snow or hail might be considered in the study or where a rain gauge might be placed e.g. north, southeast, or west sides, roof or floor, etc.
Achievable	This opens up broader questions including the size of the project and the time and funding available. Areas to consider here are how realistic is it to successfully complete the objective based on other constraints, for example does project funding run out after 6 months? Does the researcher have the skills needed for the project? Is there appropriate equipment available to measure the rainfall? Is there an appropriate area on the site for monitoring rainfall? Has the project been agreed with relevant stakeholders? Does the planned timescale fit with project deadlines?
Relevant	Relevance is considered in relation to an aim, so this would need to be evaluated based on why the project is taking place, the overall aims or values of the organisation or setting in which the researcher is based, and, of course ensuring that it is not already repeating work when data might be available elsewhere (see section 1.2.3).
Time-bound	The 12-month aspect gives a precise time point when it's clear that readings should stop, although the overall project plan should also include extra time to analyse the results and report on them.

1.3.2 How are aims and objective different?

This is something that people often find quite challenging, and it's important to remember that the objectives shouldn't just repeat the aim again with perhaps slightly different wording! Objectives are steps that set out how the aim is broken down. Aims will generally be quite broad statements about the project purpose, whereas objectives give intermediate SMART goals for the project, dividing it into measurable steps.

If we use a non-conservation example for this, let's assume that you would like to play the piano, but you have no previous experience (Figure 1.6). Your aim could be that you want to be able to play Beethoven's *Moonlight Sonata*. With no experience, you can't expect to do this straight away without mistakes. Your objectives are needed to help break down how you're going to get to that overall aim. They might include learning to read music, buying the required kit, perhaps finding a piano teacher or some online instruction videos, and, of course, lots of practice!

Like playing music, writing good **aims** and **objectives** is something that also needs quite a lot of practice, and refinement, to make sure that they are useful.

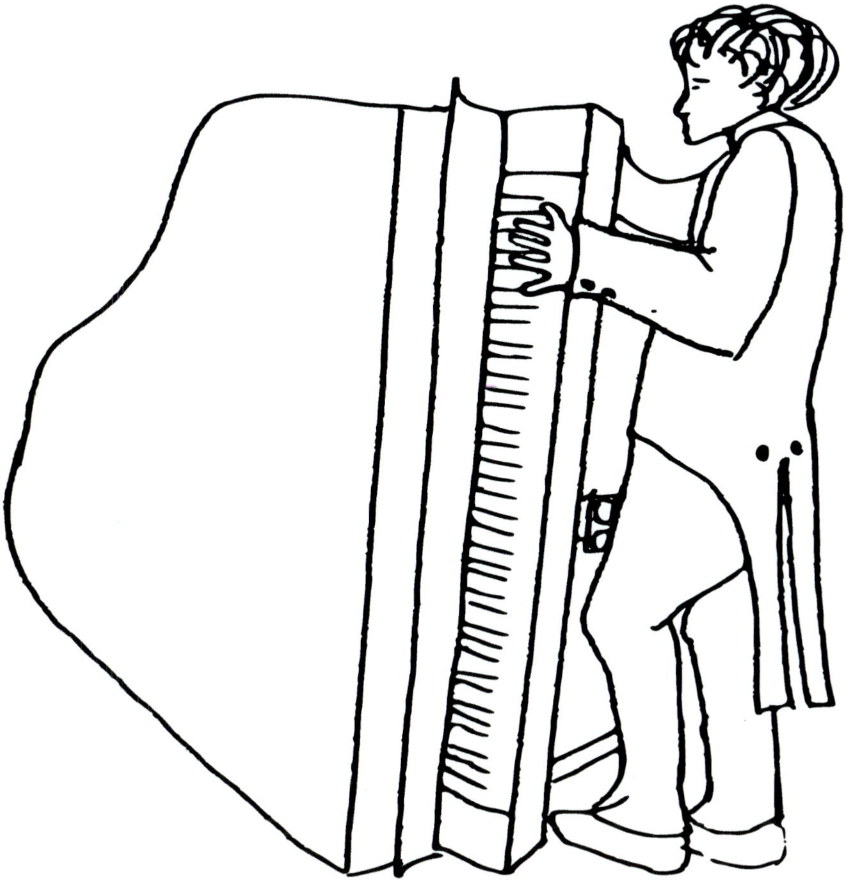

Figure 1.6 Learning to play the piano is broken down into a series of steps, just like preparing aims and objectives. Image by Philip J. A. Skipper.

These are key building blocks for your research project, so taking time to get them right is always worthwhile.

1.4 Determining a hypothesis

When the aims and objectives have been worked through, returning to the research question and refining it further as necessary, it is then time to prepare the hypothesis.

1.4.1 What is a hypothesis?

A **hypothesis** states a predicted outcome of the testing. In scientific circles, a **theory** is a hypothesis which has been tested many times and proved to be correct.

A hypothesis is developed from the potential outcomes of the research based on the research question, aims, and objectives. In order to develop a valid hypothesis, there are four terms which need to be understood.

Firstly, the hypothesis itself. This should state what the researcher expects to find from the experiment. For example, our hypothesis might be that ethanol is an effective cleaning agent for surface dirt on PMMA (which stands for polymethyl methacrylate, a modern synthetic plastic). When developing a hypothesis, it is important to avoid **confirmation bias**; this is where you design the **hypothesis** in such a way that you will always be proved correct. A (slightly ridiculous) hypothesis that 'ethanol is a more effective cleaning agent for surface dirt than acetone on PMMA' would be biased towards ethanol being more effective, as we know that acetone will dissolve the PMMA.

It's conventional to also have a null hypothesis, which states the opposite outcome compared to the hypothesis. Following on from the initial example, a null hypothesis for that project would be that ethanol is not an effective cleaning agent for PMMA.

Box 1.4 gives some other examples of hypotheses and their corresponding null hypotheses to demonstrate how this could be applied across a range of different materials and research projects.

Box 1.4 Examples of hypotheses and their corresponding null hypotheses

Hypothesis: The intensity of the green pigment used on Victorian green wallpapers directly correlates with the levels of arsenic present.

Null hypothesis: The intensity of the green pigment used on Victorian green wallpapers does not relate to the arsenic concentration.

Hypothesis: Increasing ultraviolet light levels causes more rapid fading of textiles.

Null hypothesis: Increasing ultraviolet light levels does not cause more rapid fading of textiles.

Hypothesis: Addition of zinc oxide nanoparticles to silicate-based stone consolidants reduces biofouling.

Null hypothesis: The presence of zinc oxide nanoparticles in silicate-based stone consolidants has no impact on biofouling.

The text within the hypothesis also establishes the variables used in the testing. In research, variables are the aspects that can change. For a hypothesis, we would often see a **dependent variable** and an **independent variable**. The independent variable is the part that has an overall effect on the dependent variable. In the example, ethanol would be our independent variable; this is the cause of some kind of impact on the dependent variable (the dirt on the plastic). With ethanol as the independent variable, this implies that the researcher is perhaps planning to vary the ethanol concentration in the testing process or the exposure time. For the testing to be reliable and **reproducible**, the surface of all the test plastic would ideally need to have the same level of dirt across all of it, and all the testing needs to be carried out on the same (chemically identical) plastic for a repeatable testing method. The researcher would also need to compare ethanol to another commonly used cleaning technique for PMMA. This is what makes the difference between a scientific method and a **case study**, since case studies are often not repeatable and may carry too many variables to give a robust answer to some research questions. However, there are times when case studies are the most appropriate approach to investigate a particular research question.

Look back at box 1.4 and try to identify which is the dependent variable and which is the independent variable for each of the hypotheses. Answers are at the end of the chapter.

A poorly defined hypothesis might be too vague so that it becomes impossible to prove. For example, a poor hypothesis might state that the cleaning of leather stops mould growth. This is a poor hypothesis because many factors could result in mould growth; the mould growth can depend on environmental conditions, the cleaning method used, the type of leather, previous exposure to mould spores, and many other possibilities. A stronger hypothesis could be that cleaning leather with a particular solvent inhibits regrowth of mould, while the null hypothesis states that cleaning leather with the solvent has no impact on the regrowth of mould. The statement now explores one cleaning method (the independent variable), and we could further test this under a range of environmental conditions to see whether the hypothesis is correct.

1.4.2 What is the value of a hypothesis in research?

Hypotheses are not always present in conservation research projects, but they are a standard science-based concept, and conservators would find this a useful addition to their experimental planning. The creation of a hypothesis will make the purpose of the research even clearer through making it obvious what is being changed and what is not. This can be very useful when developing the overall method, as a common pitfall when developing and planning a project is to try and change too many things at once. This then makes it very difficult to interpret when writing about the results.

The hypothesis also reduces the risk of bias through allowing the researcher to evaluate the results against both the hypothesis and null hypothesis to see where the results are the best fit.

1.4.3 How do aims and objectives relate to the hypothesis?

All these aspects have a common theme; they should be written clearly and concisely and are designed to inform the approach to the research project. They sit within the overarching framework of scientific method and experimental design. Aims and objectives are vital to a successful research project and would be expected to be included within the project output. Aims are a more general statement of purpose and how that might be achieved, with objectives giving more specific details. A hypothesis and null hypothesis are an approach informed by standard scientific practice in experimental design. Their purpose is to give a statement of what is expected – or not – within the results. The hypothesis therefore is a tool to extend the aims and objectives and give the researcher a way of specifically testing and evaluating their results against a defined outcome. If we revisit the rainfall at Lincoln Cathedral[1] example (box 1.5), this helps illustrate how they work together to set the scene for the project. There are some practice examples in the exercises at the end of the chapter and in the supplementary information on the website to help with applying this knowledge to other examples.

Box 1.5 Example of how aims, objectives, and hypothesis fit together as part of project design

An aim for the project could be: Determine whether annual rainfall levels and seasonal rainfall patterns at Lincoln Cathedral, UK, have changed in the past 100 years.

Objectives could be: (1) To compile data for historic rainfall using existing records [for the purposes of this exercise, we'll assume those records exist]; (2) Measure levels of daily rainfall over the course of 12 months at Lincoln Cathedral; (3) Compare and analyse rainfall data to investigate whether there are any changes in rainfall.

The hypothesis for this project would therefore be a predictor of the potential outcome and could read: Annual levels of rainfall and seasonal rain patterns have changed at Lincoln Cathedral over the past 100 years. The null hypothesis would therefore be the opposite or negative of that: Annual levels of rainfall and seasonal rain patterns have NOT changed at Lincoln Cathedral over the past 100 years.

1.5 Summary

This chapter was designed to help you consider the initial stages of your experimental design and planning to refine your key research question(s) and start to break a project down into manageable sections of work using a scientific methodology. We have discussed how to create and refine a research question and how this is developed into the aims, objectives, hypothesis, and null hypothesis. We have also discussed the importance of situating your work within the knowledge of other people's work in the field through reviewing literature and an understanding of what other data resources might be available. Taking time to produce a detailed initial design means that you can move into the next stages of planning with a clear focus and understanding of your 'why?' – the justification and background for the work that you are intending to do.

Regular review and revisiting of this initial design stage is helpful as you move through the project. Aims and objectives may need to be adjusted as your experiments progress. Science is unpredictable, and sometimes, unexpected results can mean your initial project plan needs some changes because of the new information gained. It is important to use these as a project basis and to regularly revisit them throughout the project to help you track your focus and manage the project, but they can be viewed as adaptable if necessary. Major changes may suggest that you also need alterations to a hypothesis and null hypothesis. At the early stages of the project, though, aims and objectives, along with your hypothesis, will help you then move into the next stages of experimental design, considering what you are going to look at and how you are going to do it in Chapters 2 and 3.

Exercises

Exercise 1.1 The researcher is exploring the use of different microscopy techniques to identify the type of wood as part of a technical examination of 15 Japanese Shinto statues. Write a potential aim and two or three objectives which could be used.

Exercise 1.2 You have been commissioned to test a new solvent which the manufacturers have called Auxilium Conservare.[2] The manufacturer claims it is a more sustainable alternative to dichloromethane (DCM). Write a hypothesis and null hypothesis for the research.

Exercise 1.3 What key words might you use to find previous peer-reviewed research on archaeological coin collections from the Indian subcontinent? Try them in a search engine of your choice to see what information is available.

Notes

1 This example is based on the weather at the time of writing, and is not based on any real project to the best of our knowledge.

2 This is a fictional solvent name.

2 How?

Experimental design methodology

Introduction

When designing scientific experiments, the methods used as part of that experimental design are tools that generate information. In order to choose which tools are the right ones for the job, you need to view this working backwards. Understanding the types of data needed to meet the aim of the study means that the researcher can then evaluate a range of possible tools and decide the best ways to obtain that data. Jumping in too quickly and doing testing without proper planning could mean that the data generated is something that doesn't help answer the initial aims and objectives of the study. It certainly causes delays and uses up your valuable time. Potentially, there may be additional costs to repeat the work too.

This chapter is all about the *how* of experimental design because we are considering *how* that information, or data, is generated. This can be considered in two parts: firstly, the data that you need to answer the research question, and secondly, the types of methods you could use to generate that data.

How you are doing the research builds on the understanding of *why* you are carrying out the research. Planning experiments should be a circular approach. You may decide at this stage that the research you'd planned isn't feasible. Perhaps you need access to an analytical technique that you aren't familiar with, and you would not be able to carry out the project as you hoped. At this point, you might wish to consider whether to revisit the research aims. A small change to the original aim may still mean that you can carry out the research but in a slightly different way.

Alternatively, you may need to consider collaboration. It may be that you need to collaborate because you do not have experience with or access to equipment needed to produce the type of data that you have identified in the planning stage. Rather than rewriting the experimental plan, collaborating with others who offer that skillset may be the way to overcome the problem. This is not the only benefit of collaborative working – working with others on a research project can add additional input and new perspectives, which often enhances the process. Research can sometimes feel frustrating when things are not going right, and a collaborator is also a useful part of a support network so that you have someone to talk to when it's not progressing as you'd hoped. It can also allow you to access new funding sources, because some funding bodies have specific calls for multidisciplinary work.

DOI: 10.4324/9781003397137-4

This chapter talks about general principles and types of data with examples and does not aim to be a catalogue of all available techniques or an instruction manual for different equipment types. For delving into the technical detail of how to carry out different tests, the resources suggested in the reading list will provide additional guidance.

2.1 Types of data

A **method** is defined as what the researcher is doing to carry out their research and is a descriptive term. Microscopy, **chemical spot tests**, and all the other techniques discussed in this chapter are examples of methods. **'Methodology'** as a term should be used differently; it is the descriptive method in combination with the justification for why that method is being carried out. An example of a **methodology** would be visual analysis including microscopy to determine the weave structure of a textile. The **method** (microscopy) is being used for a reason (determining weave pattern). When publishing your work, the format and style of the **output** are likely to make a difference to whether you need to discuss the methods or the methodologies (or both!) in writing. However, the **methodology** is the area you should define in the planning stages of a project to make sure that you are integrating both how and why together in decision-making.

When considering different data collection methods, one aspect to explore is the type or category of data needed. **Quantitative** data generates information that has a **numerical value**. This numerical value can cover a very broad range of different types of data. In the case of conservation, numerical data could be the number of silverfish caught in pest traps across a museum, or it could be a measurement of the percentage composition for a metal alloy or perhaps the number of threads per cm in a warp and weft if we return to our textile example from the previous paragraph.

Qualitative data is descriptive data. Again, the types of data that can be collected under this category are very variable. A questionnaire to conservators asking them why they have chosen a particular type of adhesive for a specific material can use open questioning through a free text comment box and allows the respondent to write in the reason for their choice. Microscopic examination of a wooden chest may show small traces of paint, and the information that the paint is red or green falls into a qualitative category. Elemental analysis is often qualitative when it indicates major versus minor elements but not percentage composition. Sometimes, qualitative data can still be analysed in a quantitative way; for example, categorising responses to a survey, where 80% of conservators stated that their chosen adhesive was non-yellowing. Alternatively, additional information can be gained through some analytical techniques. A **spectrophotometer** could be used to characterise the colour of the red or green paint, which could relate it to a colour reference or use numbers to describe the colour quantitatively. That colour reference would still give the qualitative information about it being red but also add quantitative data giving the specific shade of red. As described here, some information gathered through analysis has the potential to generate both quantitative and qualitative information.

The type of data you have feeds into how it can be analysed and presented, which we discuss further in part II, Chapters 4 and 5. Both quantitative and qualitative data can be useful for a research project, but the type of data required is ultimately determined by the research question (with accompanying aims and objectives) which has been produced in the first stage of the experimental design process. In order to choose an appropriate technique to produce the data, it's important to understand the potential options you have available. This is covered in the next section.

2.2 Types of technique

This section has been subdivided into types of technique used for the research based on the broad categories of information that would be collected through that method. Because of the wide variety of different techniques and types of materials they may be suitable for, this section aims to give an overview of some of the different methods commonly found in conservation programmes and larger museums with a heritage science laboratory.

2.2.1 Visual techniques

One of the first things that any conservator does when starting an object or site-based project is to look at the object or site. **Observation** is crucial to our work in many ways, and part of the conservation process of documenting an object involves careful recording of the overall item and specific details, perhaps through photography or drawings (digital or hand drawn), so the basic principles of observation and what it can be used for are likely to already be quite familiar to you.

Observation is also a useful tool to build into any scientific investigation. Although some specific types of observation are discussed here as examples, it is important to recognise that observation also forms a part of many other techniques, such as deciding which area on an object is suitable for sampling or measurement.

In general, observation-based techniques generate qualitative information rather than quantitative information. The visual appearance of a feather after cleaning trials with different methods could be viewed using the naked eye, using a **microscope**, or for a highly magnified image, using a **scanning electron microscope (SEM)**. In this example, the researcher would not generate numerical data but would be assessing for signs of damage, for perhaps traces of the cleaning agent left behind, and for removal of dirt (or lack of!) using the different methods. This type of investigation can still be set up using an experimental methodology, but it relies on the interpretation of the researcher to decide, based on available evidence, which is the most suitable cleaning technique. This leads to potential areas of bias, where (if cleaning is inconsistent across the surface) an image may show a 'dirtier' area from one technique and a 'cleaner' area from another technique, which would not be a fair comparison of the two different methods. However, it is possible to come up with an agreed scale of cleanliness with visual (image-based) examples, which would allow you to score the efficacy of the cleaning technique and would help with analysis. This sort of analysis is discussed further in section 2, Chapter 4.

Observation techniques may use different types of light source to give different or enhanced amounts of information. Microscopes are available in a range of different types with different functions. A **stereo microscope** generally uses light from above or the side to visualise features which do not transmit light through them, such as the surface of a coin, or can be placed over a sizeable object to examine part of it. A **compound light microscope** (also known as a **research microscope**) is used to examine a sample mounded on a slide and tends to offer a higher magnification range than a stereo microscope. **Polarising light microscopes** use as illumination **polarised light**, still within the visible spectrum but filtered to transmit light waves all oriented in the same direction (under normal circumstances, light waves are oriented at all the different angles that are possible). This polarising filter, in addition to an analyser in the path of the light transmitted by the sample on the slide, allows identification of minerals, wood, fibres, and pigments through the ability of those materials to modify the light rays passing through them when they are mounted on a microscope slide. You can identify features linked to manufacturing such as stress patterns in plastics, which are not visible under standard visible light conditions.

Examination in **ultraviolet (UV) light**, **X-rays** and **infrared (IR)** light are other **observation** techniques. These light sources form part of the **electromagnetic spectrum** (Figure 2.1). These categories relate to different wavelengths of light, with the shorter wavelength of UV light referring to the fact that the distance between each light wave's peak to the next peak is shorter than found in visible light. Shorter wavelengths of light mean an increasing amount of energy is contained within the light. When objects are exposed to UV light long term, we often see an increased rate of damage because the energy from UV light is transferred to the object. In simple terms, this energy leads to increased chemical reactions, which we see as faster deterioration, such as fading, of light-sensitive objects (which is discussed in more detail in Book 4, *Colour, Colour Measurement and Colour Change*, in this series). Because of this sensitivity, conservators try to avoid exposure to light as much as possible, especially light with shorter wavelengths like UV. As an analytical tool, however, a brief exposure to UV may give researchers valuable information about the object, and such a brief exposure is not expected to cause any noticeable damage. Further discussion about the ethics of research and evaluating the risks to objects will take place in Chapter 3.

The human eye can only see light which falls into the visible spectrum shown on Figure 2.1. Although UV, X-ray, and IR are not visible to the naked eye, we can use equipment that is capable of recording in those wavelengths to generate an image, or sometimes, exposure to these light sources will produce visual data that we see within the visible light spectrum. A UV light may be used to help determine the difference between ivory and a synthetic alternative through the presence or absence of fluorescence. Some materials, such as madder (pigment), absorb the UV light and then emit it as light in the visible spectrum, a property known as UV-induced visible fluorescence. X-rays are also a light-based technique, but their even shorter

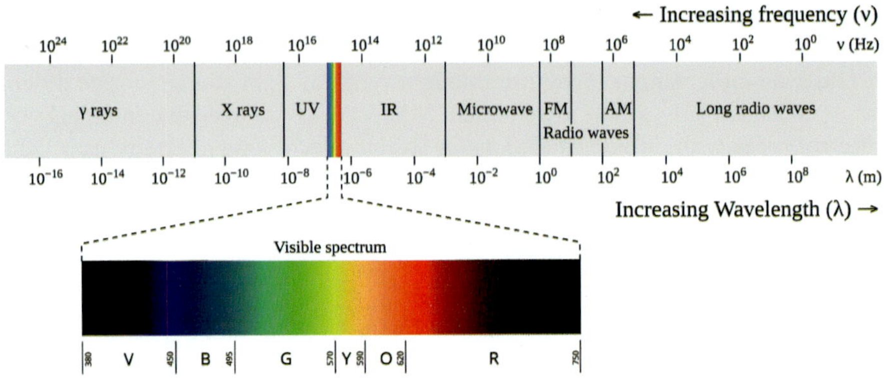

Figure 2.1 Electromagnetic radiation spectrum showing how UV, IR, and X-rays relate to visible light. Image credit Philip Ronan, CC BY-SA 2.5 <https://creativecommons.org/licenses/by-sa/2.5>, via Wikimedia Commons (unmodified).

wavelength means they pass through certain materials and can penetrate through items such as taxidermy, ceramics, or statuary to render visible their internal structures or repairs such as metal dowels. This technique produces an image which can be related to the object's actual appearance.

A broad spectrum of electromagnetic radiation can be used to visualise different aspects of an object. **Multispectral imaging (MSI)** is one example of this, in which light across a broad spectral range is used to take multiple images or a continuous scan is made through the different wavelengths. This is, again, a technique generating qualitative data. Different types of ink will become more visible on documents or drawings under different wavelengths, and MSI will help to distinguish between them.

As part of a recent exercise, we prepared drawings using a range of different inks then overpainted them. Figure 2.2 shows three of the pictures before overpainting, and underneath is the corresponding image taken by MSI, using a Foster + Freeman Crime-lite® Imager, which was originally designed as a forensic evidence photography system. This illustrates how some information is visible at different light wavelengths; because IR light has a longer wavelength than visible light, it can pass through some paints, depending on the pigments, and reveal underdrawings or paint which is carbon based. It's also a good analogy for scientific experimentation. Many students identified the cat as a potato when we carried out the exercise in class, as the legs were drawn using a blue ballpoint pen, which does not reflect IR light, while the head was slightly masked by the presence of titanium white pigment. We also see a carbon black stripe on the 'Hello World' image, which masks the underdrawing due to the interaction between carbon in the paint and IR light. Often, several techniques are needed in science to get the full picture, since incomplete data can lead to drawing incorrect conclusions. In our in-class exercise with the students, we also look at the paintings with **reflectance transformation**

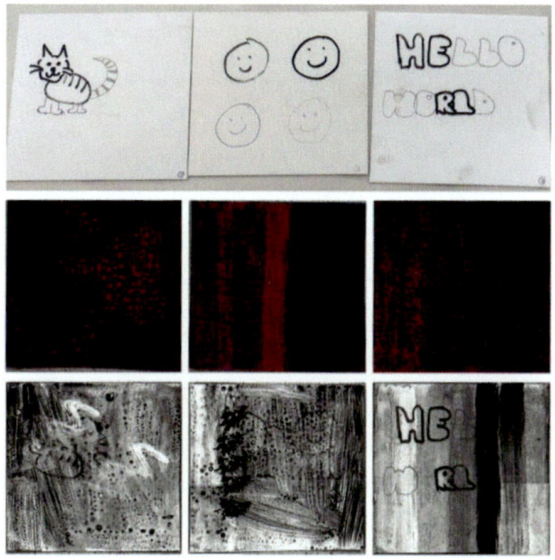

Figure 2.2 Top, three drawings prepared with pencil, biro, and two different black pens which were then overpainted. Middle, after overpainting. Bottom, the same drawings visualised by multispectral imager, halogen light (IR source) with 850 nm filter. Each lower image was assembled from 4 separate images due to restrictions to the area that the imager can visualise at one time. Image by Lynda and Philip J. A. Skipper.

imaging (RTI), which picks up the pressure marks from the biro and fills in the gaps. Sometimes, though, it can be difficult to identify where the gaps are, if you don't know what you are looking for.

Other observation techniques can include digital manipulation of images, such as is found in **photogrammetry** and reflectance transformation imaging (RTI). These techniques allow creation of a 3-dimensional virtual image, which can then be manipulated on screen to highlight or enhance features based on virtual adjustments to lighting and positioning.

Although visual information can be useful, it may be subjective, and interpretation can rely on the observation skills of the viewer, with the potential for introduction of bias based on what area the researcher is looking at or what they are looking for. Relying on personal interpretation of visual data alone can be a risky strategy for ensuring valid, reliable information, but it should not be dismissed as a research methodology because of this; when used appropriately, with awareness of the limitations, it can be a very useful part of an overall research plan.

It is not always the case that these types of observation techniques involve *only* qualitative information. Using a microscope, it is possible to examine growth of micro-organisms on a surface and generate quantifiable data through counting the number of microorganisms present or through determining percentage coverage. In fact, percentage coverage can be used for a range of different topics, such as dust coverage on a slide when assessing dust levels in an area. This type of quantitative data from observation is discussed further in the next section.

2.2.2 Measurements

Taking measurements to generate quantifiable data is a process that most conservators and conservation students will be very familiar with. Typical measurements taken of an object might include size and weight. These all give numerical information, and even something like length and weight could be considered part of a scientific experiment with an appropriate research question, such as whether arrowheads from England were of a standardised size and weight in the 1500s. Measurements are useful because they give data which can be analysed in a variety of ways through graphs, charts, and statistics, building objective evidence that can be assessed as part of a scientific methodology. How this information can be analysed and presented is discussed in section 2.

A wide range of techniques may be used to give a measurement. A spectrophotometer can be used to give a colour specification, which can show changes in surface colour over time through fading, or the size of supporting dowels in a stone statue may be measured through taking **X-rays** and then making measurements on the resulting images.

At a microscopic level, **microfading** techniques are being used to understand fading behaviours of objects as a function of colour change over time. Microfading involves the use of a very high-intensity light in a small area, not visible to the naked eye, and measures the rate of fading of the spot due to light exposure using a **spectrophotometer**, such as the example given in box 2.1. This may be used to understand behaviour of a range of different materials and to improve understanding of the impact of display conditions on their aesthetic values. We can also use microscopy to measure pigment size, which aids identification.

Box 2.1

Chan et al. (2022) used microfading to test colour stability of materials used on eight versions of The Scream *by Edvard Munch (1863–1944). A 0.5 mm test spot was used on a variety of colours, with colour measurements made every 2 seconds for 6 minutes, for a total light exposure of 0.5 Mlux (megalux). Results interpreted in terms of museum lighting levels varied from 3,000 lux hours per year to 30,000 lux hours per year before a colour change could be measured on the spot. This information about light sensitivity allowed the museum to tailor light levels, opening hours, and object rotation plans in its exhibition space to meet the needs of each artwork.*

Other properties that can be measured might include strength or hardness testing to help understanding of material properties. Measurement is not restricted to objects and heritage materials but may be carried out on potential repair materials to check compatibility with an original surface.

Measurement of desalination rates is also a common technique for conservators working with salt-contaminated materials, because it is often necessary to track desalination of an object over time to check when the desalination process has been successfully completed (the evidence is that no more salt can be extracted).

Simply measuring desalination of the object or another measurement for recording purposes would not be enough for it to be considered a scientific study, though. It is important to remember that the scientific methodology discussed in the first chapter needs to be used to structure and justify the testing. However, it may be that previously recorded information, such as time taken to desalinate a group of archaeological iron objects, could then be used as data to feed into a study on desalination techniques without needing to repeat experimentation. This is one reason it is useful to record information about work that's been done, even if it's not required for a specific research project now – this data can be revisited retrospectively and used as part of a dataset for future research.

These measurements may be used to give quantitative data to reinforce or enhance visual, qualitative indications. The researcher carrying out testing to evaluate the impact of different conservation cleaning techniques could say that a parchment 'looks cleaner' when using one technique compared to another, but someone reading an article about this might question how reliable that information is. Through taking colour measurements with a spectrophotometer of surface colour before and after cleaning, the researcher could move beyond a descriptive interpretation and demonstrate numerically that one technique caused a larger reduction in soiling than the other.

2.2.3 Chemical spot testing

Chemical spot testing investigates the presence or absence of a specific component. It can be quite a useful tool because, when performed correctly, it gives a clear visual indication of a result. It would usually generate **qualitative** data, i.e. positive or negative. Some tests may also give an indication of concentration (**quantitative**). An example is a commercial test strip used to detect the presence of copper. After adding a combination of chemicals, the test strip will change colour in the presence of copper. Nitric acid (HNO_3) can be added to filter paper and pressed onto the possible copper-containing surface. After the filter paper is removed, rubeanic acid in industrial denatured alcohol can be used as an indicator chemical, and we would expect to see a green colour if the result is positive. If there is no colour change, the result is negative. The use of spot testing is not limited to metals. Other examples include spot tests for protein, starch, and cellulose nitrate, which can help to distinguish between different types of adhesives or coatings.

Some commercially available test strips offer a semi-quantifiable option through providing a colour chart with equivalent concentrations. Generally, a strong colour would indicate a higher concentration of copper than a pale colour based on those charts, such as shown in Figure 2.3. Even existing positive and negative tests have the potential to be made more quantifiable, if required, though enhancements to the standard method. An example of this is given in box 2.2.

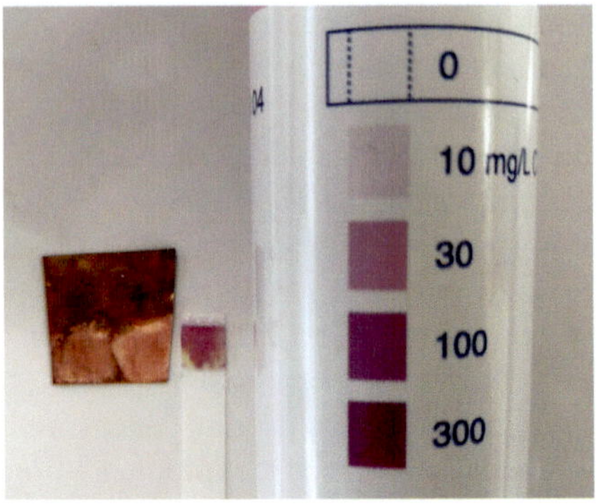

Figure 2.3 Spot test using a Quantofix test kit, carried out on a piece of copper. Left, copper foil, the tested areas appearing much brighter than the rest. Centre, test strip showing a dark-pink colour (positive). Right, manufacturer's colour scale for semi-quantification of concentration. Image by Lynda K. Skipper.

Box 2.2

Concentration of chlorides in salt extracted from building stone.

Soluble salt tests have long been used to look for the presence of different types of soluble salts in buildings. Chlorides are a common component of these salts from sources such as salt water and pollutants. The basic test involves adding silver nitrate and nitric acid to the salt dissolved in water and visually assessing whether the solution has become cloudy. Lynda collaborated with one of the students to improve the methodology using a semi-quantifiable approach, demonstrating that the level of cloudiness of the solution was related to the quantity of chlorides. We showed that comparing the solution colour to a set of visual references allows a rough quantification of the result (Skipper and Rubinstein, 2018). Although our aim with this project was to produce an on-site method for chloride testing, giving a consistent result with limited resources, use of a device such as a spectrometer to measure the turbidity of the solution would give a more precise reading.

This **spot test technique** is generally quite quick and low cost, with relatively few resources needed to complete the test. Some manufacturers sell test kits, and for others, the researcher would need to source the required chemicals. This may require investment in suitable chemical storage and in other health and safety protection for the user, such as fume extraction.

Sometimes results may be ambiguous, and **control samples (or controls)** are needed for comparison, since colour change can be subjective. Controls give known results, which allows the researcher to cross-check their observations. As with visual techniques, relying on the researcher's visual acuity to detect changes can introduce bias or subjectivity. There may sometimes be issues with the source material. A test for copper would not show positive results on a silver-plated item, even if it has copper underneath the silver, unless it is tested in an area where the silver-plated upper layer has worn away. Where chemical tests are being carried out on objects, care must be taken over the position of the test in case the test chemicals cause damage or surface change. Ideally, the potential for damage would be established through first performing a test on a sacrificial material before deciding whether to go ahead with an object-based test.

2.2.4 Elemental composition

Instrumental analysis such as elemental composition or spectral-based determination of chemical structures tends to be much more expensive to perform but can also give more in-depth information. Generally, the researcher needs a higher level of training specific to that technique to run the analysis and to interpret the results. Purchasing equipment may be out of reach of many private conservation companies and smaller museums, and maintenance and servicing add ongoing costs. That does not mean that a conservator should exclude these techniques from their experimental design though. Where the technique is the most appropriate, conservators can consider whether a collaborative approach could work as discussed in the introduction to this chapter, perhaps working with universities or large museums with access to the equipment required.

Elemental composition, at its simplest, gives a list of the chemical elements which that material contains. Chemical elements might include sodium (Na), iron (Fe), and carbon (C). Using multiple spot tests for individual components could generate something resembling a list of elements in that material but would involve repeated testing and be very time consuming. Using analytical equipment to take that reading helps to speed up the process and to reduce the risk of elements being missed. Two techniques often mentioned in conservation literature are **X-ray fluorescence (XRF)** and **scanning electron microscopy with energy dispersive X-ray spectroscopy (SEM-EDX)**: both use energy to excite **electrons** within the material. With XRF and SEM-EDX, the associated energy is unique to that element. In most cases, processing software will automatically assign elements.

Combining the detector with the SEM allows researchers to look at tiny amounts of sample and detect the elements within it. XRF tends to be applied to larger areas, though readings can take place without needing a sample if a hand-held version is used. Although neither technique is destructive, since the sample will still exist after testing, removal of a sample for the SEM examination may be ethically unwelcome – and it is generally impractical to try and replace a pinprick-sized sample once testing is completed. Depending on the setup of the instrument used, it may be possible to get quantifiable measurements, such as the percentage concentrations of metallic **elements** in an alloy, through prior calibration of the equipment.

Understanding what that list of elements *means* is something that takes time and practice. Where there are multiple layers of materials, some techniques will detect elements in all the layers as one list, which makes interpretation even more challenging. A simple metal item with a clear percentage of gold, for example, can be relatively easily categorised into the type of gold based on the other elements present. An oil painting, with perhaps some surface dirt (which will also be detected), multiple pigments, and a binder, with three layers of paint, will produce a much more complex list for the researcher to analyse if they are hoping to determine what pigments are present. Sometimes, though, the question may be more straightforward. Checking whether toxic metals such as arsenic or mercury used as pesticides are present on taxidermy can be done using XRF, and the results will give a clear indication of the main health and safety concerns that item might present from inorganic pesticides. There are limitations to the level of detection – low amounts may not be picked up – and EDX and XRF cannot detect lighter elements. If the taxidermy in our example had been treated with historic organic pesticides, they would not be detected by XRF or EDX. An alternative sampling technique is needed if organic materials are likely to be present.

2.2.5 Chemical structure determination

Spectral analysis techniques such as **Fourier transform infrared spectroscopy (FTIR)** and **Raman spectroscopy** can be used to give information about chemical bonding within a material. In those cases, light is used to excite bonds, and the energy emissions are detected and converted into a series of spectral peaks or troughs. Through comparing readings for an unknown substance to known information about materials, often via an existing library of spectral data, it is possible to gain information about the chemical structure of the material being analysed. Analysis can be carried out using hand-held equipment that can be transported to a site, or on samples which are removed from an object, depending on the type of equipment chosen.

Similarly to elemental composition, this technique can help to answer questions about what materials are made from and how they are degrading. Changes in chemical bonding due to the ageing process can be seen through shifts in the spectral peaks – relating to a change in bonding, such as increased cross-linking. Research looking at cellulose acetate and cellulose nitrate movie films has demonstrated a change in spectral peak ratios as deterioration takes place, giving a way of determining which are the most degraded films, even if differences are not at the point of being visible by observation alone. Other uses may include preventive conservation, such as looking at the distribution of deposits of oils or glycols and glycerol from haze machines used while filming at heritage sites.

These techniques can show similar drawbacks to those discussed under elemental analysis. Although these techniques can detect **organic** (carbon-based) bonds, where there are no bond vibrations present at these wavelengths (some **inorganic** materials), these components may be missed. Assigning bonds does not necessarily tell the researcher how those bonds are joined together to form the overall material, and mixtures can give complex results which are more challenging to interpret. Some instruments require removal of a sample.

Gas chromatography – mass spectrometry (GC-MS) partially addresses some of these limitations, because the sample is separated out into its different components for analysis, but the sample is destroyed in the process. It can give accurate results, such as determining the length of carbon chains in a mixture of hydrocarbons. The sample preparation and analysis of the peaks tend to be complex, and more training and experience are needed to use the equipment than the methods previously mentioned. We used GC-MS to identify unknown chemicals at a National Trust property in the UK. They had been left in containers (some unlabelled) by the former owner of the property and formed part of the collection. However, some containers were degrading, and the potential hazard to the public and the volunteer guides was unknown. For one unlabelled bottle, FTIR produced a 91% match to the hydrocarbon white spirit (not high enough to be conclusive). GC-MS was able to confirm the presence of hydrocarbons, specifically carbon chains from 9–17 carbon atoms in length. This meant it was likely a kerosene, which was commonly used in the UK as a fuel for small lanterns or stoves (Ireland, Skipper and Elie, 2014). There are a range of different variants of GC-MS (such as Py-GC-MS used for the analysis of Asian lacquers), but these are all specialist techniques which we will not explore further here.

X-ray diffraction (XRD) can be used to discover both chemical structures and elemental composition of crystalline materials. It has a broad range of potential uses in conservation, which include identification of materials and their corrosion products. Due to the complexity of the analysis, requiring specialist operators and high-cost equipment, the technique is less commonly used in conservation at present. It's often the case that a conservator can get enough data from other sources so that this more specialist technique is not seen as necessary to answer their research question.

A summary of the most commonly used analytical techniques at present and the types of data they can provide is given in Table 2.1.

2.3 Artificial ageing

Artificial ageing is sometimes known as **accelerated ageing**, as it is a technique used to age materials, usually through exposure to very high levels of light, relative humidity (RH), and/or temperature, to make the natural ageing process of that material progress at a more rapid rate. It is often used as a basis for other experimental testing. It is not a technique that can be used to generate a valid, reliable result on its own, but it needs to be paired with some type of analysis to measure the impact of the **artificial ageing** technique on the material.

Unsurprisingly, this is usually carried out on either **surrogate material** for ethical reasons, which is used to mimic effects of ageing on historic items, or on **conservation materials** which are being tested prior to use. Feller (1994) suggested that conservation-grade materials can be divided into different classes, with class A having a lifespan of greater than 100 years. If we were to invent a new material now for consolidating textiles, to check its stability before use, we would need to

Table 2.1 Summary of some commonly used analytical techniques in conservation and the types of data they can provide (shown by an x in the relevant section).

Technique	*Results*	*Qualitative*	*Quantitative*
Light microscopy	Magnified images, measurements, UV light sources can provide additional information.	x	x
Polarising light microscopy	Magnified images of individual paint components, fibres, dust particles etc.	x	Limited
X-radiography	Images through the whole depth of the object, information on the density of components and hidden internal details, can identify past repairs.	x	Limited
XRF	Elemental analysis. Not useful for organic objects but very useful for identifying metal alloys etc.	x	x
MSI	Imaging at different wavelengths, can reveal hidden detail depending on the materials in the object. Fluorescence can reveal past repairs.	x	Limited
Spectrophotometry	Measures colour accurately, can be used to calculate changes in colour over time or aid colour matching for conservation work.		x
RTI	Uses photography under raking light to reveal hidden details, for example in painted murals, engravings which have been mostly worn off, and manufacturing details.	x	
Photogrammetry	Multiple images are used to create a 3D model of the original object. The greater the number of images, the more detail. Good for documenting, accessibility, can be used on larger objects (up to full buildings) to plan conservation work.	x	Limited
SEM-EDX	Provides high-resolution magnified images; EDX produces elemental analysis similar to XRF, including lighter elements. May require samples to be removed from the object.	x	x
FTIR	Provides information on organic materials based on library spectra. Mixed materials can be difficult to identify. Some types require sample removal.	x	x

wait for over 100 years to find out if it could be classified as a class A material. This is clearly not a practical approach, and so artificially ageing materials to check their long-term behaviour is the current best method available for assessing their suitability and then not using the ones that prove unsuitable. Through using very high light levels, it would be possible to check the effects of 100 years of ageing in just over a month (see box 2.3).

Box 2.3 Light ageing calculations

Objects in museums typically have a light budget for visible light exposure. That can be expressed as 50 lux of UV-free light for sensitive materials, for example, or as a yearly exposure. The annual budget in lux-hours does vary slightly between different organisations, but for the purposes of this example, we will assume a maximum of 150,000 lux hours per year for to be allowable for a highly sensitive object. This is calculated using the following equation:

Lux hours = light level in lux multiplied by time exposed in hours

An object exposed for 50 lux over 12 months of the year, 8 hours per day, receives 2,920 hours of exposure (365 days x 8 hours = 2,920 hours). Lux hours are then calculated as 50 lux x 2,920 hours = 146,000 lux hours per year. This is roughly the yearly level suggested for very sensitive material, of 150,000 lux hours, so we will use this to make rounding easier.

Returning to the artificial ageing, we use the same equation. If we assume that the object is exposed to light at 150,000 lux hours per year, after 50 years display ageing the object will have been exposed to 7.5 million lux hours (that's 150,000 x 50 years).

We know the total amount of light needed for the artificial ageing, so now, we need to speed this up. We can substitute this value in the lux hours equation above to give:

$$7{,}500{,}000 = \text{light level} \times \text{time}.$$

Let's assume that the ageing chamber produces 20,000 lux, and we need to work out the time the item needs to be in the chamber to generate the equivalent of 50 years ageing. Rearranging the equation gives us:

$$7{,}500{,}000 / 20{,}000 = \text{time in hours} = 375 \text{ hours}$$

light exposure at 20000 lux.

Assuming that the light is running for the whole day, that's the equivalent of 50 years display in just under 16 days (calculated by 375 hours divided by 24, the number of hours in a day).

Ageing can also be carried out in the dark, using relative humidity (RH) and/or temperature to increase the rate of reactions, or via a chamber with exposure to high pollutant gases or even a combination of factors. It's also possible to introduce different types of ageing through mechanical stress or wear measurements or by daily cycles of light exposure and darkness or daily variations in temperature and RH to mimic heating or air conditioning that runs only during a museum's open hours. It is not surprising that the more variables there are, the more complex the calculations become.

Where RH is constant, the **Arrhenius equation** can be used for temperature-based ageing (box 2.4). It is commonly stated in the conservation and chemistry literature (and by conservation lecturers!)[1] that rates of reaction roughly double for every 10-degree increase in temperature. One might expect based on this, that temperatures of 80°C would age a material 64 times faster than if it were in a museum storeroom kept at 20°C. That's assuming it doubles (x 2) at 30°C then doubles again (2 x 2 = 4) at 40°C, doubling again (x 2 x 2 x 2 = 8) at 50°C and so on. However, this is a substantial oversimplification of the process.

Box 2.4 The Arrhenius equation for calculating ageing

$$k = Ae\frac{-E_a}{RT}$$

k is the rate constant
A is the pre-exponential factor
e is Euler's constant
E_a is the activation energy
R is the gas constant (8.314kJ mol - 1)
T is the temperature (in K).

A conservator looking at this for the first time will likely have already identified some issues in calculating this. With light, the variables are something that conservators can measure and fit into the equation themselves. For this reaction, much more complex factors come into play, which conservators will not necessarily have the ability or the technology to measure. However, some information is often available from other studies of materials, and the conservator need not try to derive all the figures from scratch.

Luxford and Thickett (2011) calculated k/A as the overall rate at different temperatures, using the published activation energy for silk (50 kJ mol^{-1}). They argue that A is unlikely to change significantly at the temperatures used (20 to 80°C), and therefore, a comparison between reaction rates at the different temperatures is a valid approach. They calculated that silk at 80°C aged approximately 33 times faster than silk at 20°C when RH was kept constant. This is half the rate at which the 'doubling for every 10°C increase in temperature' approximation would suggest. Different materials with different activation energies would show different rates.

T_g, the **glass transition temperature**, must be taken into account when choosing suitable ageing conditions. If artificially ageing an adhesive that reaches T_g at 35°C, the impact of ageing at 40°C would simply be to soften or perhaps even melt the adhesive. It would not mimic the normal ageing process at room temperature in cool climates. In this example, using light and/or high RH would be more appropriate.

Another disadvantage of this technique is that it does not accurately reflect the range of conditions materials might be exposed to in reality, such as fluctuating temperatures and RH levels and the impact of interactions between different environmental deterioration processes. Because of this, it can really only be used to draw indications of how a material might behave when aged.

A range of international standards is available for artificial ageing protocols which set out clear parameters for the ageing process. In general, using a standard protocol is advisable because it enables your work to be cross-compared to that of other researchers using the same protocol. It is also an international standard for a reason – this protocol will have been agreed on by a group of experts working in the field as being a suitable method to use for that particular material type. They usually also set out the methods required to determine the effects of the ageing, which will help your experimental design.

2.4 Mixed-methodology approach

Sometimes, multiple methods have the potential to provide data that can answer your research question. While this might mean that decision-making in relation to experimental design becomes a bit more complicated, it is actually something that should be viewed as a benefit.

In those scenarios, it's useful to consider a range of factors to help with decision-making around which method or methods are required for the research.

- Which methods might give you answers to the research question?
- Are there any restrictions on sampling, e.g. the quantity or size of samples that can be taken or areas which can be tested?
- Which methods do you have access to or could commission if required?
- Is sufficient expertise available for interpreting results?
- Do the methods fit within the constraints?

In this instance, constraints can be a range of factors but might include access, health and safety, sustainability, budget, time scale, and ethical issues. These are further explained in Table 2.2 and are listed alphabetically, not in priority order. Can you think of any other constraints which might have an impact on your project?

We like to view this decision-making process as a triangle (Figure 2.4). At the top of the triangle are all the potential methods. Although it is possible there might only be one method at this point, it's more probable that there will be four or five, or perhaps even more, that then need filtering down as part of the experimental design process to ensure they are suitable.

Table 2.2 Constraints to consider during experimental design.

Type of constraint	*Ways in which it might impact the decision-making*
Access to the item/site	Is scaffolding required (which also impacts cost, project management regarding timing of access)? Is the object off site or in store, and how might access be obtained?
Ethics of testing	Ethical testing will be considered further in the next chapter. To summarise this here briefly, ethical issues may include removal of sample(s) or the risk of damage to the item.
Health and safety	Is sufficient protection in place for the researcher and others (e.g. museum visitors or staff) while the research is carried out? This might include extraction of chemicals, manual handling, or exposure to X-rays or UV light. Some objects are also at risk from handling, e.g. deteriorating plastics.
Sustainability	Researchers have a responsibility to consider how their work might impact the climate through use of chemicals and carbon emissions generated. Where one approach is more sustainable than another but gives a similar result, then this should be a preferred solution.
Time scale	If results are needed for an exhibition, a conference, or a student dissertation, these will have fixed deadlines. Setting up an experiment that takes 6 months when the deadline is in 4–6 months will clearly not be appropriate since it leaves no time to evaluate the results.

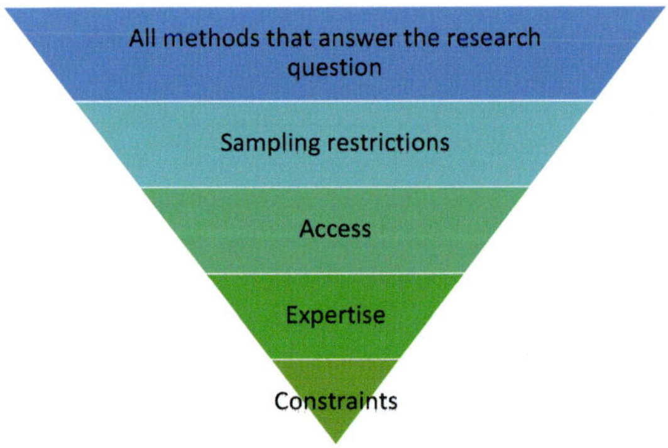

Figure 2.4 Stages of a research methodology decision-making triangle. Image by Lynda K. Skipper.

Consideration of any sampling restrictions is likely to reduce the numbers further. For example, some methods may need more sample than is available – or perhaps all methods need to be non-destructive, with no removal of sample at all. Obviously, if the researcher does not have access to the method, that reduces the number of possibilities further. If there is not expertise available to interpret the analytical results, then that is clearly not going to be beneficial for understanding the outcome of the study. By this point, the number of available techniques is likely to have been reduced further, and the researcher can now consider any additional constraints specific to the project and to the organisation or the time available.

This may leave perhaps one or two methods that are suitable for the project and is a way of refining the approach to research design and structuring the decision-making process around choosing a methodology. An example of how this might work in practice is shown in box 2.5.

Box 2.5 Worked example of methodology decision-making

Research question: Carry out analysis of a historic mortar to enable matching with a historic mortar recipe

Potential methods:

- *Microscopy (grain size and sharpness/inorganic components)*
- *Thermogravimetry (inorganic components)*
- *XRD (inorganic components)*
- *XRF (elemental analysis)*
- *FTIR (organic components)*
- *Mass spectroscopy (organic components)*

Sampling restrictions: Heritage site, requires removal of mortar sample. Maximum acceptable sample size 10g; needs to be unadulterated for analysis, so it must be possible to remove the outer layers to analyse the inner core.

Access: The site will be scaffolded for access to mortar areas, but there is a narrow time frame for analysis as recipe needed before reconstruction starts.

Expertise: Microscopy, XRF and FTIR are available, but no in-house access to other techniques or their interpretation skill set.

Constraints:

1. *Budget: which methods would be most suitable for the funds available? XRD will give all the answers that thermogravimetic analysis would, both give more detailed information than XRF, so outsourcing should be included in the budget. Providing the samples are suitable, FTIR will give similar information to mass spectroscopy, but worth ensuring that the budget has enough flexibility to include outsourcing mass spectroscopy if possible.*
2. *Time: need to know the sample turnaround time for outsourced jobs.*

3. *Sample size: in this case, all techniques require low sample sizes, so 10g should be more than sufficient.*
4. *Knowledge: no in-house experience in interpreting thermogravimetry, XRD, or mass spectroscopy.*

Looking at the research question using this decision-making process helps to highlight potential problems. For example, with constraint 4, knowledge, we are now aware that we need to make sure that the results from outsourced analysis need to have a report with a full interpretation of the results, not just an answer. We might also consider using multiple techniques to confirm the results, although then, we could run into higher constraints regarding the budget.

Even though the list of different options may have been reduced through this approach, it is important to recognise the benefits of having a mixed-methodology approach to your work. A literature review, as discussed in Chapter 1, is always necessary as background to your project. In addition, a literature review should also form part of your overall methodology and will need to be included within the publication or output of the work. This allows you to place your work within the field and sets the context for what others have done. You would also use the literature review to aid your experimental design process, informing the decision-making for your methodology by understanding how other researchers have approached their experimental design for similar projects or in the same field. Literature reviews will also enhance your understanding and interpretation of the results through cross-comparing your findings to others.

It is likely, then, that you will be carrying out at least one experimental technique alongside a literature review. That's a useful starting point, as ideally, the hypothesis might then be supported by data from the experimental analysis and through a review of literature. It is possible, though, that data from just one technique might generate information that could be interpreted in a couple of different ways, and so it may not give an accurate answer to the research question. To help ensure that results are reliable and that data is valid, using a third – or more – research methodology can help to enhance your confidence in the conclusions. This concept is shown in Figure 2.5.

Multiple methodologies can also help build information in areas where they are not directly overlapping. For example, one technique might offer partial information, and a different technique might also offer partial information. Taken together, those techniques give a more complete picture – rather like adding pieces of a

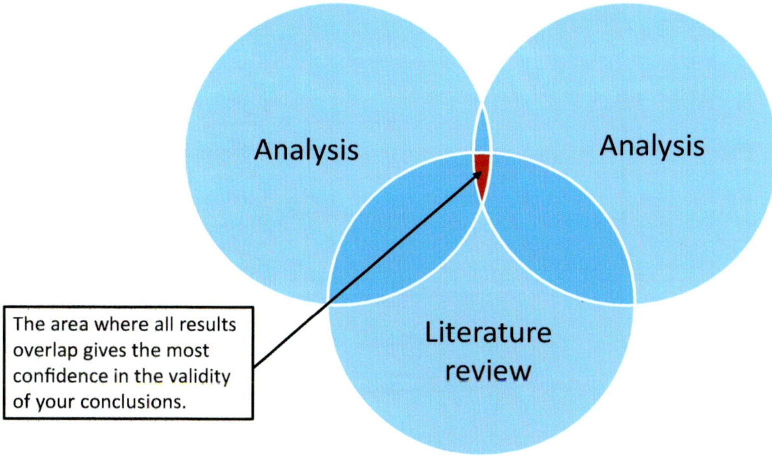

Figure 2.5 Mixed methodologies give more reliable data. Each circle represents a technique. For increasing confidence that you have reliable and valid data, all the results should point to the same area, shown in red on the schematic. Image by Lynda K. Skipper.

jigsaw puzzle together to build the overall finished image. Box 2.6 gives an example to illustrate these concepts.

Box 2.6

When viewing the FTIR data in Figure 2.6, the position of the peaks suggest that the two main components being analysed are the same.

Visually, Quinacridone Red and Quinacridone Violet look different, as the name suggests, with one pigment being violet and one being red. This colour difference can be measured with a spectrophotometer applied to paint containing these and observed using PLM. They are both synthetic organic pigments consisting of carbon, oxygen, hydrogen, and nitrogen. However, using FTIR, the spectra for the two pigments look almost identical. This is because they are polymorphs – in other words, they are the same compound but with a slightly different arrangement in their crystal structures due to differences in hydrogen bonding. In this example, either FTIR data or PLM analysis alone would have potentially led to an incorrect conclusion. Visual data alongside literature about the pigments helps to give a reason for the analytical results.

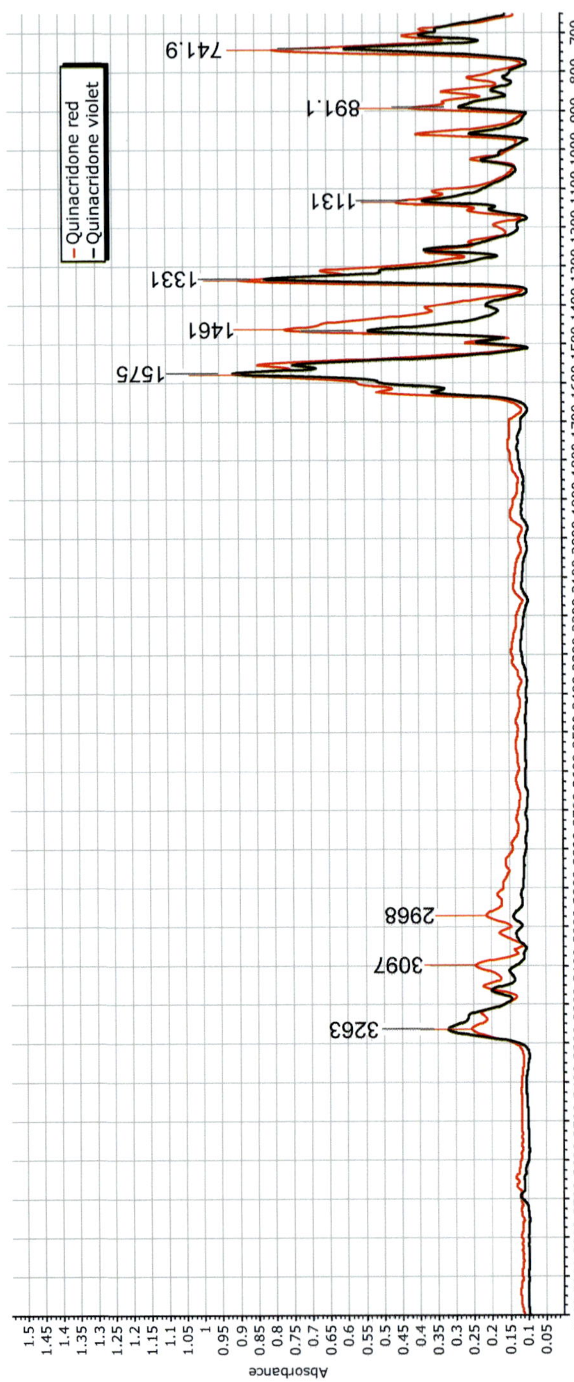

Figure 2.6 FTIR data from Quinacridone Red and Violet pigments. Image by Philip J. A. Skipper, graph drawn in Spectragryph and modified for publication in Inkscape.

2.5 Summary

This chapter was designed to give an overview of the types of data that could be gathered using different types of techniques. This overview of the different options is not written to tell the researcher which one is 'best' for their project, because each project is different, and each experiment will involve the collection of different types of data. This data needs to feed into the research question, aims, and objectives and provides the method by which the researcher can accept the hypothesis or the null hypothesis. Sometimes, this can be done through using just one method, but it often needs a combination of methods. Through thinking about and planning the reasoning behind choosing a method, it becomes a methodology, because the researcher is able to justify why the method they are using is suitable. This decision-making process about the analysis and methodology should be integrated into the design phase of the project so that the researcher has a clear decision about the types of information needed to answer the research question *before* beginning the experimental phase – it is too late to start thinking about analysis of the data afterwards!

Other elements that also feed into planning and design are the conservation ethics of testing. They have been touched on briefly here in respect to particular techniques, but this is a topic which will be explored in more detail in the next chapter.

Exercises

2.1 Experimental design exercise.

- a Think about what information would be useful to obtain for an object you are currently working on or one that you have seen in a museum and found interesting.
- b Use information from this chapter to identify techniques which could be used to give you this information.
- c Identify potential constraints and brainstorm some ideas about what you could do to overcome these problems.

2.2 Given the method or methods that you chose for exercise 2.1, try to turn the method into a methodology by noting down the explanation for why that technique is useful in the research.

2.3 What type of methods might be suitable for answering these research questions?

- a What is the composition of a 19th-century bronze statue?
- b What types of inks are present on a historic parchment?
- c What type of plastic is a modern artwork made from?
- d Is it safe to strip the old white paint in this building, or does it contain lead?

Note

1 Sample size of 2. The sample size would be an important point when reporting the answers given in questionnaires.

3 What?

Ethical sampling and scientific rigour

Introduction

This chapter completes the why?, how?, and what? of research for the first section of this book. For the final chapter of the three, we will be considering the scientific method approach and how this integrates with conservation values and ethics. We also explore the concept of valid and reliable research through the use of controls built into the project as part of the design phase. Humble science discusses the principle that science may be flawed and how important it is to be open to new ideas and the possibility of mistakes.

Interpreting data in science is always open to uncertainty. While we work with the best knowledge available at the time, that does not mean that our interpretation is always flawless. Making sure that everything is fully recorded through the process will help to ensure that work can be revisited and reinterpreted or that others can follow the same protocol for further investigation in the future. Having an ethics- and values-driven approach to their work and careful attention to planning and recording are not new skills for conservators to learn. These are already integrated into our conservation practice, though as with all skills, these are areas which can be learnt and do improve with practice and application. This chapter aims to illustrate what conservators need to consider when applying these skills to their scientific research to complete the experimental design process.

3.1 Ethical sampling

Conservation ethics states that actions taken by conservators should consider the tangible and intangible significance of the object, there should be a clear rationale for the decision taken, and actions should be appropriate and sustainable (AIC, 1994; Icon, 2020). Although different conservation organisations worldwide have slight variations on their ethical codes, the underlying message is the same. It's important to ask if the actions you are proposing might cause damage to the object, and if that is the case, can that damage be justified? If considering it from a risk assessment perspective, we are asking whether the potential benefit outweighs the risk. This is the underlying principle behind the idea of **ethical sampling**. Despite the name, it is not only the practice of taking a sample from an object but has broader relevance to any experimentation which involves original heritage materials.

DOI: 10.4324/9781003397137-5

Values and **significance** also come into play here. If we were to say that we have a great idea for a new cleaning technique for paintings, which we're going to test on the *Mona Lisa*, an iconic painting produced by da Vinci in the early 1500s, we would expect that suggestion to be met with strong resistance! And justifiably so, because any damage caused by carrying out tests of an otherwise untried technique on an iconic painting would have a substantive impact on its significance and values (this is not limited to financial **value** but may impact aesthetic, research, and educational value depending on the type of damage caused). That's not to say that the research would definitely cause damage, but testing on an item which has much lower significance or a surrogate object is the logical step before moving to an object which has a higher significance. This can also be considered in terms of risk and benefits. All testing on objects carries some risk, but when testing on an object with low significance, the benefits to the conservation community as a whole from this imaginary new cleaning technique could be argued to outweigh the risks to the object which is being tested. However, the first stage of ethical testing is to ask whether testing really needs to be carried out on an object at all. The review of information available, which was discussed in Chapter 1, may identify that the data already exists and may instead need collating and evaluating in a way that better explores the research question. For the purposes of the next section, though, we will assume that the researcher has decided some sampling or measurement is necessary in order to address the aims and objectives.

When carrying out experimental work, it is often preferable to avoid the use of original heritage materials entirely and carry out testing on surrogate materials instead, as mentioned in the example. Surrogate materials are those which mimic (as far as possible) the properties of the material you would like to test on. One example of a surrogate material could be the use of newly created porcelain samples to test fill materials instead of testing them on historic porcelain. This has an added benefit of being reproducible. It is easier to get multiple samples that are all the same for testing when working with new materials that have been purchased or made specifically for the experiment. The potential disadvantage is that new materials may not be chemically identical to the heritage material. They will not have been subjected to the same ageing or weathering processes and historic cleaning or repairs, and so there may be instances in which they could behave differently. Artificial ageing protocols can be used to address the question of ageing materials, as discussed in Chapter 2, but the surrogates will not have been exposed to exactly the same conditions as an object over its lifespan. A **surrogate object** refers to a whole object, whereas the term **surrogate material** tends to relate to a small piece of a substance, such as using a small square of porcelain for testing fills (surrogate material) rather than making a whole teacup (surrogate object).

Where surrogate materials are not a viable alternative, we can consider the use of non-destructive testing. A non-destructive test is one in which the experimental design involves using techniques that do not cause any change to the object or its materials. For example, a hand-held or micro-XRF for elemental analysis can be used to detect the different elements within a material (within certain limits, see Chapter 2) without needing to remove a sample. If the researcher was intending to study the composition of a historic piece of glass, using surrogate materials would be meaningless because they would not be the same composition as the glass being studied.

Some techniques are classed as non-destructive but still require a sample. Mounting a fibre on a slide to investigate the fibre types in a historic tapestry is technically non-destructive, and we can assume the conservator would have followed good conservation practice by choosing an unobtrusive area to remove that fibre from. The fibre still exists after it's been viewed under magnification. The researcher is unlikely to re-attach it to the tapestry, though, and in this case, it would be considered a sample. If stored somewhere, the sample may be at risk from dissociation (one of the agents of deterioration) where information linking it to that tapestry could be lost. At that point, it becomes a fibre on a slide with no explanation and is not meaningful for future study. It is at risk of being disposed of due to the lack of context.

If we continue thinking about this tapestry, another approach could be to use an already loose fibre, perhaps from the back of the tapestry, or one which is loose in the storage box for the object. Using something that is already detached means that no damage occurs to the object itself, but the researcher cannot be fully confident about whether it is from the main tapestry, a fibre from a previous repair, or even a fibre from a previous researcher's clothing! This approach may still be valid, but the results should be interpreted with caution, and the source of the fibre should be documented.

In contrast, some techniques such as GC-MS can involve total loss of the sample in question. The gas chromatography part is likely to require the sample to be dissolved or vapourised in order to separate it into its component parts. This would be a destructive testing process, but it generally uses a very small amount of sample, potentially an amount that is barely visible unless examined under magnification. Deciding whether some damage might be acceptable in order to carry out the research is discussed in an open-access document called 'Ethical Sampling Guidance' (Icon, 2019).

Ethical sampling guidance is designed to help researchers navigate the decision-making process involved with using historic objects in their research work. It is relevant to techniques which involve a process of sampling, whether that is destructive or non-destructive. Although the term 'object' is used here, that should be viewed broadly to cover any historic materials, whether they are part of built heritage or any other historic item or collection.

If we assume that sampling from the object is necessary, then we can move on to consider the process for this. It is likely the researcher will need to consult with **stakeholders** before any work takes place in order to obtain permission to carry out the sampling. The first step for being able to gain informed consent is for the researcher to have a thorough understanding and plan for the experimental design which can be presented to the relevant stakeholders. It is also absolutely key that the need for interventive sampling can be justified through a clear research question and clearly articulated aims and objectives for the work. Returning to the idea of risk and benefit, the researcher must be able to show that the benefits of the research outweigh the risk to that object.

In relation to the impact on the item itself, the researcher should have planned:

- How many samples might be needed
- Where those samples would be taken from
- The size of samples needed
- How the samples will be taken
- How the samples will be analysed

During the discussion, it is possible that the stakeholders may ask for plans to be modified or sample numbers reduced, or they may refuse access. If we assume that the sampling can go ahead in some format, further discussions and decision-making are needed around the data from the study. The researcher should have an approved plan for **dissemination** of the results. Sometimes, research is carried out for internal use in an organisation and may form part of the object record but not be made more widely accessible. At other times, dissemination via academic publications and/or public engagement with the research may help to demonstrate the potential benefits (impact) of the work to the organisation or stakeholders.

There should also be a clear plan for the samples after research has been completed. Will they be returned by a researcher, and how will they be stored? Similarly, discussions should take place about **raw data** and whether that should be shared with the owners of the object. In some cases, it may be necessary to discuss ownership or copyright of data. Timing of the project is also important so that **stakeholders** have a clear understanding of any timelines so that they know what to expect from the research and when. Even in museums with facilities for in-house analysis, it is necessary to associate samples and reports on findings for the long term.

Projects do not always go as planned, so it's also important to discuss the 'what if' scenarios. Perhaps the analysis may have failed for one sample; are more samples an option, or is it not possible? Timelines may change due to a wide variety of factors, from the researcher being ill to equipment breaking down, or simply other commitments, meaning the project does not progress at the expected speed. Good communication with all parties is absolutely essential to make sure that changes can be accommodated if needed.

It is vital to record in detail the discussions with any stakeholders so that **documentation** of everything that has been agreed is in place before any interventive work on the object starts. A signed agreement helps to protect all parties in the research process. At times, a pilot study may be needed to generate initial data or provide a proof of principle for a larger study. This concept is discussed further in section 3.2.

3.2 Pilot studies

A **pilot study** aims to carry out a smaller data collection exercise, which can then inform design of a larger or main study. It would not be used to test the hypothesis in itself but to check the feasibility of an approach or methodology. Sometimes,

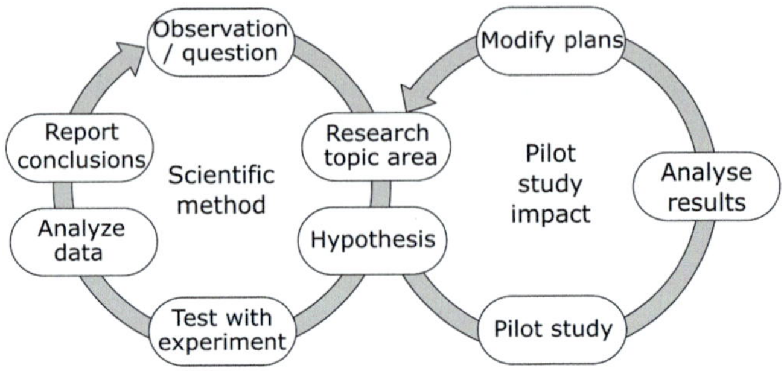

Figure 3.1 Scientific method schematic modified to add the impact of a pilot study on the design phase of the project. Original image by Efbrazil, CC BY-SA 4.0 <https://creativecommons.org/licenses/by-sa/4.0>, via Wikimedia Commons, modified by Philip J. A. Skipper.

carrying out a pilot study can help the researcher to decide how many samples are needed or to give an idea of the timescale required for the study. At other times, it may be useful to make sure that the preferred analytical technique does work successfully for the planned study. Where a pilot study is used as part of the scientific method, we can consider the way this information works within the project framework by modifying the scientific method image originally shown in Figure 1.1, Chapter 1, to give the schematic shown in Figure 3.1.

3.3 The importance of controls and repeats within experimental design

Positive and negative controls are used to demonstrate that experimental results are valid through checking that the testing process itself is working correctly. A **positive control** would be a material that is expected to give a positive answer. For example, if using FTIR to examine an adhesive sample which the researcher believes is likely to be fish glue, running a sample of known fish glue as a comparison would act as a positive control. In this case, the positive control would also help to show that the experiment – and the equipment – is working as expected if the two spectra are a good match for each other. It is not always straightforward, because if the fish glue has chemically changed on ageing, there may be problems with making a direct comparison if running a new fish glue sample as a positive control.

The purpose of a negative control is the opposite of a positive control. It is used to generate a result that is not positive. Again, it gives the researcher a cross-check that their experimental process is working correctly. **False positives** occur when a result is positive, but it should not be. By carrying out a negative control, it can help to validate results through demonstrating that the experiment has been correctly performed. False positives may be caused by a range of different factors, including equipment malfunction, cross-contamination of chemicals or equipment, or through user error (e.g. mistakes in following a sequence of steps to carry out the experiment). In the case of the FTIR example, the negative control would typically be carried out by performing a background check. This background reading is subtracted from the spectra automatically. Although we would hope that the researcher has checked that the instrument is suitably clean before use, if any traces of chemicals from a previous run remain, this removes them from the data. However, researchers cannot always rely on the equipment to perform suitable negative controls for them, and they must consider how both negative and positive controls can be built into the experiment alongside testing. Box 3.1 gives an example of how this might be applied.

False negatives may also occur in experimental research, which is a result appearing negative when it should not be. Similarly to a false positive, a false negative may occur through problems with the equipment or chemicals, incorrect technique, or perhaps missing a step in the experimental method.

If the controls do not work as expected, the researcher will need to re-evaluate the experimental design and try to work out what's caused the problem, as this means any results from the measurements would also be invalid. Ideally, the researcher would use controls to check the correct operation of the testing *before* carrying out testing on the object or other samples, because this prevents them wasting time gathering data they cannot use and wasting samples that might not be replaceable.

Box 3.1

If we return to the chemical spot test for copper used as an example in Chapter 2, section 2.2.3, pause for a moment and consider what you think suitable positive and negative controls might be before reading on.

As a positive control, using copper would give a known material that should give a positive result. This will demonstrate that the test is working, and it also gives a comparison to the expected colour of a positive test.

As a negative control, the researcher will need to find a material that is known not to contain copper. A pure sample of a different metal, such as aluminium, would be one option. An additional negative control could be to use a material with no metal at all, and this would show that any colour change observed is not due to the presence of interactions between the test chemicals or cross-contamination of the materials used.

If the positive control and the sample both give the same colour, it indicates a positive result. However, if the negative control also turns the same colour, it shows that there is a problem with the test. This illustrates the importance of considering both positive and negative controls as part of experimentation!

When carrying out tests on an object, it may not be appropriate or possible to carry out repeated tests. This would need to be considered based on the **ethical sampling principles** discussed in section 3.1. However, standard practice for most scientific experiments is to carry out multiple tests of the same thing, also known as **repeats**. The purpose of this is to ensure that the data you are getting is reliable. If someone carries out a test and bases conclusions on just one instance of the result happening, then that is quite high risk. Sometimes, data can be an outlier. **Outliers** are defined as data that is different from the standard or expected result (Figure 3.2), and they can happen for a variety of reasons. Sometimes, there has been an error in the experimental design or setup that means the result is incorrect. Sometimes, these outliers might be due to overall variability, either through the measurement method or variation in the material or testing itself. Finally, outliers can be due to some new or novel phenomenon. This is sometimes a reason research can divert from the original aims and objectives, because an unexpected result takes the research down a new and different path. If there is only one data point, though, it becomes impossible to identify these outliers. Generally, we would exclude data that is clearly generated through experimental error from the analysis, but with only one repeat of that particular data point, it becomes more difficult to identify those, and the researcher could form incorrect conclusions based on flawed data.

In the example graph drawn in Figure 3.2, if a **trend line** were drawn to include all the data points including the potential outlier, the trend line would be a lot higher on the y axis to take into account that outlier. Looking at the sketch, most people would assume the data point was flawed and choose to exclude it from the analysis, given that the other points follow a fairly clear pattern. The way to be

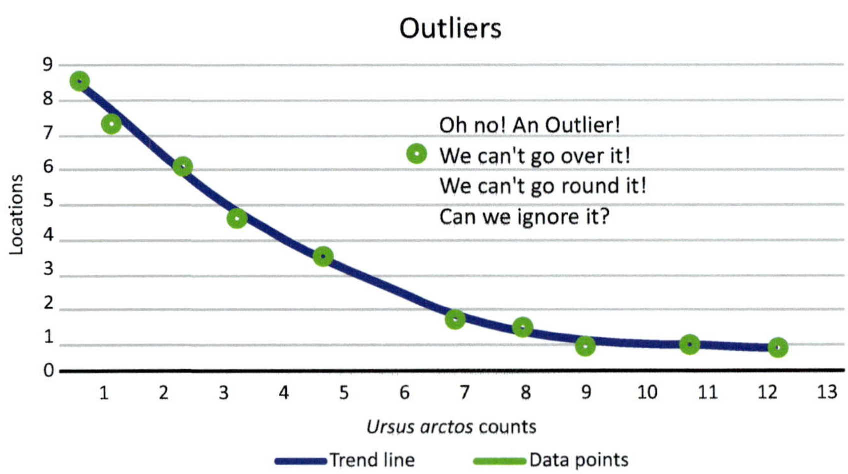

Figure 3.2 Line graph with data points joined by a trend line, currently ignoring a data point which appears to be an outlier. Image concept by Lynda K. Skipper, created by Philip J. A. Skipper, inspired by Michael Rosen.

more confident of this, though, is to take enough repeated readings at each data point that outliers can be identified more confidently.

Figure 3.3 shows some recent testing we carried out measuring the growth of algae on stone. Each block of stone was analysed on every quarter – in other words, four repeats per block – to record the colour and algae levels. The same test was repeated on four different stone blocks, giving a total of 16 data repeats. After seeding the surface with algae, the stones were placed in a glass lidded chamber and the algae grown in a climate chamber at optimal growth settings for algal species. On opening the chamber, we observed that one-quarter of one stone was much greener than the others (Figure 3.3, lower left). Close inspection showed that the stone block was slightly oversized in one corner. The larger size of the stone meant that the stone surface was touching the glass lid. This provided a point for condensation to form in the humid growth environment, and the extra water that pooled in that area led to increased growth rates of the algae. By taking the time to carefully observe the experiment, we were able to identify that this result was an outlier, and we could exclude it from the analysis.

When following a scientific methodology, three repeats are usually considered the minimum quantity of data points to allow confidence in the results. Having three repeated results which give roughly the same answer means that the data is said to be reliable. This allows the researcher to draw meaningful conclusions through minimising the level of uncertainty. There are different ways in which reliability can be considered. Measuring the same area three times gives a more reliable result for that particular reading; the researcher can be more confident that the reading is accurate and that the equipment used to take the reading has not malfunctioned. Repeating the readings on three different samples, which have all gone through the same process prior to testing, indicates that the information from that set of tests is reliable, as all have behaved in approximately the same way. Box 3.2 gives an example of planning repeated tests and controls.

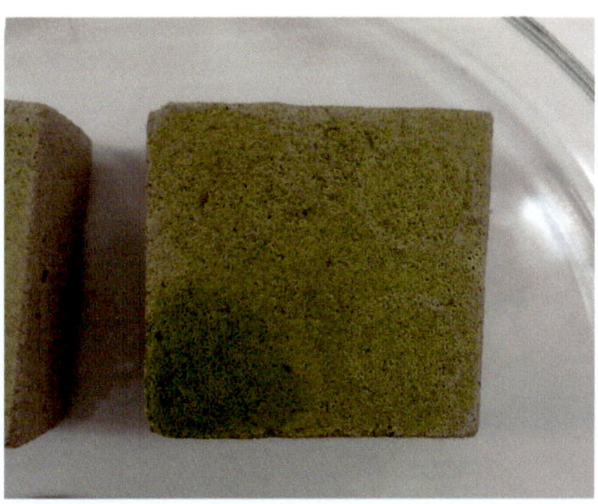

Figure 3.3 Careful observation showed that the high level of greening on the lower left corner of this stone was due to experimental error, and the data point could be excluded from the analysis of results. Image by Lynda K. Skipper.

If two out of the three data points show one result, and the third result is quite different, then the researcher needs to consider their approach. If there is clear evidence that it's an outlier because of experimental error, then that data point can be removed without concern. For example, perhaps they noticed the climate chamber for that part of the test had been switched off by another user of the space (and yes, that is also a real example from the authors' lab work),[1] and so it had not been exposed to identical conditions to the other samples. Assuming that you can then exclude one data point as an outlier, that leaves two data points. While this isn't ideal, it's generally seen as sufficient to draw a reasonable conclusion, though the preferred option would be to repeat the test to get at least one more data point. However, if you are working to a deadline, this may not be possible.

If the reason for the outlier is unclear, then the researcher can – if constrained by a deadline – discuss the results and acknowledge the uncertainty in interpretation. That can be a frustrating outcome, since it may not then prove possible to either accept or reject the hypotheses with any level of certainty. The preferred solution, of course, would be to carry out more repeats – preferably an additional three samples, though there may be constraints around cost and time which mean it is not practicable to carry out that many.

As we have already acknowledged, there are circumstances in which multiple repeats are not possible for practical and/or ethical reasons. This should be evaluated in the outputs from the study as part of the limitations. In order to give the reader confidence in their findings, authors need to make sure that they acknowledge these potential limitations and account for and discuss them within the study itself. Being only able to repeat a test once instead of the ideal minimum of three times would be a limitation of the study.

Box 3.2 An example of repeats and controls

In a project to look at the lightfastness of printer ink, there is a range of potential variables. If we assume the researcher chooses to use two different types of paper and three different printers, that gives 2 x 3 = 6 examples. When we add repeats, each data point should have three repeats to give more confidence in the results.

If we map out our repeats in table format (Table 3.1), we can see that this now gives a total of 18 samples. Each ink is repeated three times on each paper. In this case, we assume that the sample can be checked at time = 0 and can be re-used after each set of testing.

We would then need to add control samples. In this case, it would be difficult to look at the ink alone, so a control for the paper would be blank (unprinted) paper for each of the paper types. This would mean three samples for each paper type, giving an overall total of 24 samples.

Table 3.1 Counting samples needed for an example test, including repeats. Rows are shaded from clarity between the different testing sets.

	Paper 1	*Paper 2*
Ink 1	x	x
Ink 1	x	x
Ink 1	x	x
Ink 2	x	x
Ink 2	x	x
Ink 2	x	x
Ink 3	x	x
Ink 3	x	x
Ink 3	x	x

Table 3.2 Counting sample repeats needed for a more complex set of tests, including destructive testing. X shows where the sample is being tested at the different time points.

Paper 1 ink 1	*Sample number*	*Time = 0*	*Time = 1*	*Time = 5*	*Time = 10*
	1	x			
	2	x			
	3	x			
	4		x		
	5		x		
	6		x		
	7			x	
	8			x	
	9			x	
	10				x
	11				x
	12				x
Paper 2 ink 1	13	x			
	14	x			
	15	x			
	16		x		
	17		x		
	18		x		
	19			x	
	20			x	
	21			x	
	22				x
	23				x
	24				x

If some of the samples are tested destructively, such as for tensile testing, it may be necessary to add extra samples so that some can be removed at different time points to look at changes over different time periods (e.g. 1 year, 5 years, and 10 years equivalent exposure for the artificial light ageing). Re-drawing the sample table just for ink 1 on both paper types (Table 3.2) shows that 24 samples would be needed to collect three repeats of the data at each sample point; a four-fold increase in the sample number (98 in total if testing all the variations).

3.4 Effective record keeping and documentation

As conservators, we are trained from the start of our careers to keep thorough and accurate records of what actions we have taken. Those records may be used in treatment reports and attached to object records to ensure that conservators can understand what actions have been taken in the past and to inform conservation treatments in the future. This documentation process also forms part of the ethical code of practice for conservators.

When carrying out conservation science and experimentation, it is important to have the same kind of thoroughness of recording. We tend to use a handwritten lab notebook, which is updated daily, though for some projects, we prefer to keep an online record. Some notes may be quite brief, and other days may have longer notes, for example if exploring options for experimental design. Some pages may contain sketches of experimental setups, mind maps as discussed in Chapter 1 that are used for planning, or tables of results. It's always useful to have a record of your overall thoughts and planning, even if some options do not get taken forward into the main project design, because they may be useful for future projects. Other people may prefer to carry out their record keeping electronically, making sure that back-ups are made. The format and style used for a lab book is down to personal choice, but the crucial point is that it needs to exist in some format to meet ethical requirements for documentation of work.

It is important to record everything, even if the experiment did not appear to work. Your reflections on this and why it might not have worked will help you plan experiments in the future. More broadly, reflections and evaluations of what worked well and what did not, along with thoughts about what might be done differently in the future, are all aspects that can be recorded in a lab book, alongside the factual recording of what was done and on what date.

When Lynda was working in science, her lab book was countersigned weekly by the Head of Department to validate that she had carried out the work as documented; for Philip, it was a monthly review by someone in the same field but a different department. This was, in part, to show the work had been carried out on the dates stated, in case our results generated data that could be used for a patent application. In heritage science, it is less common for that level of cross-checking and countersigning to take place, but it is something that could be considered depending on the project being carried out.

When we have peer reviewed articles in the past, we have realised that methods are not always documented in the paper as thoroughly as they could be. The authors need to give enough information in papers that the method can be repeated by another researcher, and if the data is reliable, another researcher should be able to get the same results. Having detailed notes about the method in a lab book makes it easier for the author of the article to relay comprehensive information about their method to others. We cannot claim to always be perfect here; one of the authors recently spent nearly an hour trying to work out a detailed protocol for some Raman spectroscopy. It was distributed across four different electronic files because the protocol was refined over several different sessions – and each note about it contained a partial method!

Having a good record of what is planned and then a record of what has been done will improve confidence in the scientific rigour and validity of the information being presented. It helps to build trust in the reader that the work is correct.

3.5 Scientific rigour and validity of data

The phrase '**scientific rigour**' is commonly used (and we're aware that we've also used the term in earlier chapters without exploring its definition in detail), but what does it mean, and how does it relate to the context of heritage science and conservation? Allende (2004) suggests a definition of **rigour** as 'the disciplined application of reason to subjects related to knowledge and or communication'.

In other words, scientific rigour is the thorough application of everything we've been covering in this book. We've used a range of different examples of research projects as case studies to help with development of a rigorous experimental design, and some additional aspects will be explored later in this chapter. In the case of heritage, if work is published that is not sufficiently rigorous, it risks encouraging others to learn from work that is inaccurate. In worst cases, others applying the same process could cause damage to the objects and historic sites in our care, or the process may be hazardous to those that are carrying out the work.

Much of the responsibility for scientific rigour lies with the researcher. Researchers should take an ethical approach to scientific data in much the same way as they would take an ethical approach to considering which conservation treatment to apply. Although there are checks through the peer-review process of work before publishing, this process cannot cover every possibility. Obviously, falsification of data is unethical. At a lesser extreme, a researcher who deliberately chooses to ignore data which does not fit their expected outcome is not demonstrating sufficient scientific rigour. However, this can create a tension – at what point does an outlier, which we discussed under controls and repeats, become deliberate ignoring of data? This is where having enough repeats can help with making a decision about what is appropriate. It is also important to have a thorough record of what happened. If records suggest that an outlier may have had slightly different conditions for some reason – perhaps the test was performed on an unusually hot day or the equipment had a brief power failure that may have caused errors – then it is easier to justify the reason for excluding a result or set of results from the analysis. Sometimes, extra repeats are needed to increase confidence that the results are sufficiently rigorous. Aspects such as accurate data reporting in an article are something that reviewers will generally take on a trust basis. They will need to accept that the results are accurate representations of what happened, because it is unlikely that they will try to repeat them themselves as part of the peer-review process. This is also known as validity, which is often mentioned in relation to scientific rigour.

Validity is the confidence that the method(s) carried out are reliable through use of control samples and repeated experiments and suitable choice of techniques for carrying out the testing and for the analysis afterwards. When writing up your output from the study, making sure enough precise and accurate detail is given about how you did the work will give the reader more confidence in the validity of your data. As with scientific rigour, in order for results to be valid, they need to follow the scientific method and to think about the different factors which may influence this. Box 3.3 gives an example of how validity can apply in practice.

Box 3.3

In an experiment carried out to measure the efficiency of gel cleaning of photographic materials, the researcher has thought about independent variables such as which gels are being tested and the type of dirt being removed. However, it wasn't considered that the surrounding environment may have an impact. If we imagine that the repeats for one gel are placed in full sunlight, the cleaning effect may be different when compared to repeated tests of a gel (using a different mixture) which are placed on a fully shaded area of the workbench. That cleaning difference might be due to different types of gels, and the researcher could potentially use that data to draw that conclusion. However, because they have not considered the environment, they cannot rule out possibilities that the difference might be due to the temperature increasing reaction rates (from infrared heating of the gels in the sunlight) or the influence of visible or UV light or that the gel may dry out more rapidly in sun and be less effective. Because of the different variables which are uncontrolled, this experiment would not be valid.

Understanding limitations of the work forms part of a broader concept known as **humble science**, which we'll explore in the next section.

3.6 Humble science

People sometimes make the assumption that science is always right. Well, it isn't. This may seem a rather blunt response, but it is critical for understanding of science that everyone understands that it is not perfect. Conclusions are drawn based on the data available at the time. Sometimes, the data is flawed; sometimes, the interpretation was mistaken; and sometimes, advances in techniques and technologies mean that results are reinterpreted in a different way because of new understanding and advancements. In science, this is one of the reasons why we distinguish between a hypothesis (an idea) and a theory (a hypothesis which has been extensively tested and [thus far] always found to hold true); we are acknowledging that it takes a lot of work to tell whether something is right or not. It can sometimes be hard for people to acknowledge their potential sources of error or bias and that their work might be wrong. When it is identified, researchers may publish a correction or retraction, depending on whether they wish the published information to be fully removed or adjusted.

Published scientific work usually goes through a fairly rigorous system of checks. We use other sources in publications, through literature review and **citations**, to show where ideas come from and what other work has been carried out in the field. We have self-imposed checks like controls and multiple repeats, which were discussed in section 3.3 of this chapter. When writing up the outputs of a project, there should be enough information provided for someone else to follow the method, repeat the work, and overall get the same results. Submitted publications are looked at by a peer-review process, which asks experts in the field to give feedback and request changes prior to any publication. Even so, this process is sometimes flawed. Arguments and findings presented in published work can still be superseded by new information, or perhaps experts may disagree over how findings are interpreted (see example in box 3.4).

Box 3.4 Lynda's initial experience of humble science

As a PhD student working in biochemistry, I was examining the rate at which a protein called neuroserpin forms polymers (long chains), because in some rare cases, a mutant form of this protein causes dementia. As part of measuring the rate of polymer formation, I needed to purify the protein after growing it in bacteria in order to study it. A colleague had found that adding glycerol in the purification stage made protein yields a bit better. It wasn't until after our paper was published, giving rates of polymerisation, that we realised the glycerol had bound to the protein and was stabilising it, and our published figures were wrong. The reason this came to light was a colleague had purified some of the same protein without glycerol and had been getting different results when they tested the protein properties. This led my research down a path of exploring whether it might be possible to stabilise the protein, which could potentially lead to future treatments. We, of course, also published a correction to the original paper. As a student, I'd already been exposed to the idea that science is fallible, but to be directly involved with the process also taught me a lot about humility and the importance of being willing to openly admit mistakes.

NASEM (2019) completed a review of repeatability and retractions across a range of research fields, though conservation and heritage did not form part of their study. They offered a range of explanations for why work may not be repeatable, including issues with statistical data and a rush to publish (due to funding pressures). Interestingly, the concept of repeatability itself is not always clear-cut, and the ways that others measure repeatability vary between different studies.

Although we have shared a science-based example from one of the authors of this book, conservators themselves are generally becoming more open and accepting of mistakes as a profession. For example, the American Institute for Conservation currently holds a mistakes session at their annual conference, where conservators can share mistakes and learn from others. Conservation research is an ongoing process of learning, and so it is not surprising that new information is often arising which helps to improve our practice. In my experience, conservators generally have a culture of sharing learning and being open to new ideas, which we need to continue to cultivate – and to appreciate.

If we return to the concept of humble science, how can we use this concept in our scientific research and experimental design? Being aware of places where errors can be introduced and thinking about how to control for them is important, as is allowing enough repeats to be reasonably confident of the accuracy of the data. Carefully designed experiments can reduce the risk of mistakes, but they can't eliminate them. Wording of reports and publications can help indicate the level of uncertainty; using 'suggests' and 'implies' instead of the more definitive 'clearly demonstrates' phrasing indicates areas where some uncertainty in interpretation might be present (Figure 3.4).

Ultimately, accepting that you probably will make mistakes in the research process is important, but don't be discouraged by this! Sharing those mistakes to help others is also a valuable part of our work.

3.6.1 Understanding and acknowledging values and the potential for bias

The scientific method implies that researchers are working objectively to establish facts, which are then assessed to reach a conclusion. But how accurate is this really? All people have their own motivations, which are usually values driven, for carrying out research in their chosen subject, and when we are approaching our research from a values-based perspective, can scientists ever be truly objective?

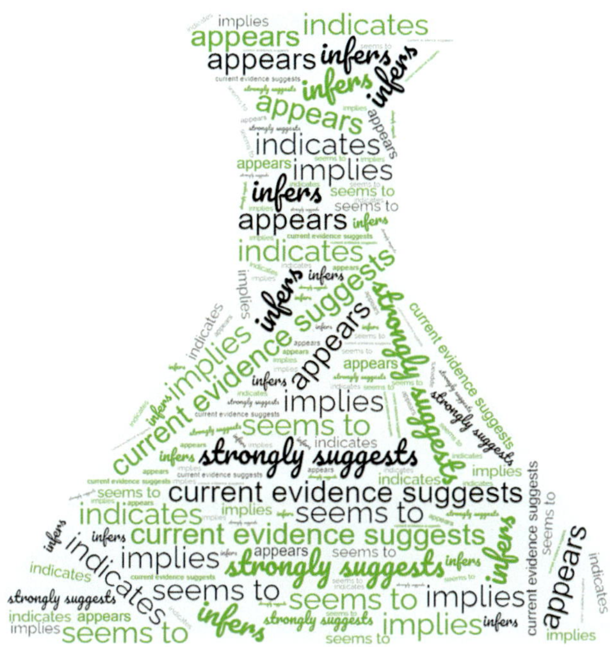

Figure 3.4 Some alternative phrases that could be used to discuss results where the outcome is uncertain or open to interpretation. Wordcloud made by Lynda K. Skipper on www.wordclouds.co.uk/ using a conical flask shape.

Noss (2007) argues that we should not see this as a flaw in our ability to follow a scientific approach but that we should think about how values-driven science works *alongside* objectivity, not in opposition to it. When our values are those of truth, fairness, and adherence to professional standards of ethics, this has the potential to improve our work rather than being detrimental to it. Although Noss was writing in the context of conservation biology, this approach also fits the approach that we would expect from a conservator in their daily work, where we acknowledge that our decisions about heritage conservation are ethics and values driven. There is no reason our scientific practice should not follow the same principles.

What is key, though, is to design our experiments in a way that allows us to test and explore that sense of truth and being open to changing our minds if that is what the results indicate. It's certainly possible that we may start the experimental work with a pre-conceived idea of what we hope might happen. In order to avoid bias, we need to be open to results that perhaps offer a different interpretation or may not be what we wanted to find. Ethically, we have to be as willing to openly publish that 'truth' instead of the results we were hoping for.

In order to avoid our values becoming biases, it is important to think about how we phrase all steps of our experimental design process. If our initial aim is to 'prove technique X is suitable', that immediately introduces potential bias. We are looking for examples where the results meet the initial aim and may exclude results that do not fit with what we were looking for. If our aim was to prove that swords from a particular time period did not contain any carbon, choosing an inappropriate analytical technique that cannot detect the presence of carbon would give a biased result – just because it cannot detect carbon does not mean that it is not there!

3.7 Summary

When we think about the 'what' of experimental design, we need to consider a range of different factors, which may make the difference between our experimental design process being a success or a ~~failure~~ learning experience.[2] This is a cyclical process, and often, the initial research question and aims and objectives may need to be revisited based on information which has come to light during the design process. No researcher is expected to get their experimental design correct the first time. Everyone goes through a process of refining and developing plans and potentially using a pilot study to develop it further. As part of the process, it's necessary to carefully consider and evaluate options for analysis and the types of data required to answer the research question. This process aims to ensure that it will be a viable project which fits the scientific method, with valid and reliable data, and shows a good standard of scientific rigour. Researchers also need to ensure that their research plans are a good fit with the ethical principles of conservation and choose approaches that are suitable for a values-driven approach to work. When mistakes are made (which is almost inevitable), it is important that the researcher reflects on that and, if work is already in the public domain, is honest about making corrections as required.

Through completing part I of this book, you have now encountered the why, how, and what of experimental design. The second part assumes that you have now gathered some data and takes you through the next steps of what to do with it, from statistical analysis to presenting results and writing reports.

Exercises

3.1 What controls would be needed if you were comparing three different types of filler for bone artefacts to see if they show any colour change over time?

3.2 A conservation student would like to test different options for cleaning pewter, but the object they are aiming to clean is historic and is not suitable for experimental testing of different processes due to potential risks to the object. How could the student overcome this problem?

Notes

1 We strongly recommend the use of 'do not touch' signs and/or signs over electrical switches if there are multiple users of a space when keeping the equipment running when you are not present is critical for the experimental design!

2 The strikethrough is deliberate – remember, your work isn't a failure if it all goes wrong! Reflect on it and learn from it so that you can use what you have learned next time.

Part II

Scientific data analysis

4 Introduction to statistics for conservators

Introduction – Why do we use statistics?

'I shall try not to use statistics as a drunken man uses lamp-posts, for support rather than for illumination'.

Attributed to Andrew Lang, anthropologist, novelist and folklorist (1844–1912).

The main reason we need statistics is the human tendency for pattern recognition and **confirmation bias**. We are very good at seeing patterns in things, such as shapes in clouds and faces on trees. It also means that we see patterns in our results which might not be there. Pattern recognition is not necessarily a negative trait; this ability can also lead to researchers exploring unexpected results through identifying patterns in data.

When we combine this tendency towards pattern recognition with the tendency to interpret information in a way which supports our prior beliefs (confirmation bias), we need a tool that can objectively verify whether what we are seeing is correct. Researchers must be willing to change their opinions based on the results of the experiment, as an underlying desire to feel validated can overrule our need to be correct. Interestingly, this is even true into research on confirmation bias! Freedman and Sears (1965), who we assume did not like the idea of confirmation bias, published a review of 14 hand-picked research reports and claimed they showed little support for confirmation bias; subsequent reviews by other researchers using a larger selection of studies showed that confirmation bias does exist. Our choice of words may also have encouraged a form of confirmation bias in you, the reader; do you trust the outcome of their study based on what we have written?

In Chapter 2, we discussed how you should consider the results and their analysis as part of the initial experimental design. Good design helps us to avoid experimental bias. If our experimental design is poor, and we are trying to justify our results by staggering through statistical test after statistical test until one gives us the answer we hoped for, then we are misusing statistics. It's also far more time efficient to select the correct test from the start.

This chapter aims to help you select a suitable statistical test based on the type of data that you will collect from your **experimental methods**. We are going to look at some of the commonly used statistical tests and talk about when to use them and

DOI: 10.4324/9781003397137-7

what they tell us. You can find worked examples of how to use them in the supplemental material.

4.1 Basics of statistics

Working out which tests to use can seem daunting, especially because there are so many tests and they use quite specialist language. You will no doubt be relieved to know that this chapter is not going to be an in-depth exploration of statistics; in this chapter we will focus on what conservators really need to know in order to use statistics appropriately. That means understanding which statistical test to use and how to interpret the results so that you are applying the statistics correctly. As long as you know which test to choose, you don't need a detailed understanding of how the equations work for all of the tests, since there are plenty of software packages that can be used as a tool for the actual calculations. For anyone who does want to extend their understanding of the equations further, in-depth information can be found in the statistics books listed under further reading.

4.1.1 Statistical terms

Before we go any deeper into which tests to use, it will be useful to define a few concepts, starting with population, sample, and case. Figure 4.1 shows how these concepts relate to each other.

Populations might represent a whole collection of objects. When we carry out a study, it is unlikely that we can test the whole population. Instead, we carry the study out on a sample and infer how the study will affect the population based on the results of the sample. If we were comparing the effects of a consolidant on Portland limestone, we couldn't test it on all the Portland limestone in the world. We would use a sample.

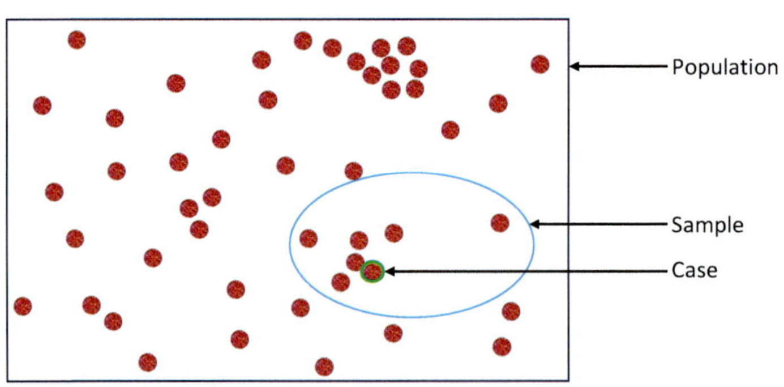

Figure 4.1 Population (black rectangle) can be studied through a sample from that population (blue outline). The individual unit in the population is a case (outlined in green). Image by Philip J. A. Skipper.

A **case** subdivides this further into an individual unit or item within the **sample,** such as analysing the composition of an individual tube of red ochre paint (case) from a collection of red ochre pigments (sample).

Descriptive statistics are used to describe the properties of the data you have gathered. They can be either qualitative or quantitative and are used to summarise your data in a manageable form. They should not be used as a test for your hypothesis because they are descriptive and will not cross-compare the data.

Inferential statistics are a group of techniques where a small sample of data from a study can be used to answer questions about the real world; amongst other things, this group of tests includes hypothesis tests, which we use to accept or reject the null hypothesis.

In this chapter, we will look at six common descriptive statistical tests and the three most common hypothesis tests.

Significant or **statistically significant** has a defined meaning in science, and the word 'significant' should never be used in reports unless you have proved it statistically. It implies that you have gathered your data and run a suitable statistical test or tests to prove that your hypothesis is true. You would need to support this claim by stating what test you have run and what the result was. Generally, this claim of significance is supported by a **probability value** or ***p*-value**. This is a measure of how probable it is for you to get this result by chance. The p-value will range between 0 and 1, and the lower the p-value, the more likely it is that you have proven your hypothesis. A p-value equal to or below 0.05 is generally considered proof that your results are not just occurring by chance (or in other words, they are significant). This number relates to the chance these results have occurred randomly; 0.05 is the equivalent of a 5-in-100 chance, 0.01 is a 1-in-100 chance, and 0.001 is a 1-in-1,000 chance. The lower the p-value, the more confidence you can have in the significance of the results.

There are two stages to statistical testing. First, you need to describe your data so you can check that the data you have is suitable for testing your hypothesis. The second stage is to infer information about the world in general from your data (inferential statistics) and test whether your hypothesis is true or not. In section 4.2, we will look at descriptive statistics and then move on to look at the hypothesis testing part of inferential statistics in section 4.3; but first, we need to think about what the different types of data are so that we can pick the correct tests for our data.

4.1.2 Types of data

When you are designing your experiments, you will need to think about the type of data you hope to generate, as different types of data determine which statistical tests you can carry out.

There are four types of data: nominal, ordinal, interval, and ratio. Nominal is the least complex type of data, and ratio is the most complex and precise. As such, we can do less with nominal data than we can with ratio data.

Data is **nominal** when you categorise your data by labelling the **samples** in mutually exclusive groups, for example categorising ceramics by manufacturer (for example Wedgewood, Hornsea, or Denby) or shape or sculptures by stone type and design. If you think about these categories, you can see that there is no specific order between the categories; it doesn't matter which one comes first. You could look at different shapes by manufacturer or different potteries by shape.

Data is **ordinal** when you have a category and a rank to your data, so for example, if you were carrying out a condition survey on paintings in a collection then one category could be watercolours, and the ranks could be unacceptable, poor, fair, and good. You could assign a numerical value to the ranks for analysis, but there is nothing between the ranks or anything to tell you how far apart they are (for example, good might be closer to fair than fair is to poor).

With **interval** data, it is possible to categorise it, apply a rank, and infer equal intervals between neighbouring cases, but there is no true zero point. A good example of this would be temperature measured in degrees Celsius or Fahrenheit; if you are taking temperature readings (your result is described as its rank) in a room (sample) at half-hour intervals, you can generally infer what the temperature change was between those intervals, and it can be positive or negative. If we were measuring temperature in Kelvin (and here, it would be wrong to write 'degrees Kelvin' by custom and practice), our results would not be interval data, as the Kelvin scale stops at 0 in contrast to the Celsius scale, which means that the Kelvin scale leads to ratio data.

Ratio data is similar to interval data, but it has a true zero, as it never goes into negative figures. If you are taking RH readings in a room, you can infer what the RH would have been between the readings, but RH does not go into negative figures, so it has a true zero.

Generally, interval and ratio data would be quantitative, and ordinal and nominal data would be qualitative. Have a look back to section 2.1 and see if you can work out why.

4.2 Descriptive statistics

Descriptive statistics means statistical methods used to describe or summarise data. This includes everything from graphs and charts through to formulae for averaging your samples or to look at whether your data is skewed. We won't cover graphs and charts in this chapter, as they are covered in Chapter 5, but we will look at some of the more common descriptive statistics. It is important to remember that descriptive statistics are designed to work within a sample, so you could look at the average of a sample or how much the cases within a sample vary from the average. What you cannot do with descriptive statistics is use them to compare between two samples. That is when you would use inferential statistics.

You may be familiar with terms such as 'mean' (also known as 'average'), 'median', and 'mode', which we talk about in the next section. These are descriptive statistics, which are 'measures of central tendency'. In other words, they allow you to simplify the results so you can present a single number from the whole sample. When you use this simplified number, you are stripping away all of the variability in the sample. However, there are times when it is useful to show this as well, which leads us to another category of descriptive statistics: 'measures of variability'. This includes tests such as the range and standard deviation, and this is what you are looking at when you see results presented in a 10 ± 0.05 style format; the measure of central tendency is 10, and 0.05 is the measure of variability. Measures of variability also allow us to check whether the data we have is skewed to one side or misshapen, which might make it unsuitable for statistically testing the hypothesis. This descriptive stage is an important step which shouldn't be neglected.

4.2.1 Measures of central tendency

Rather than looking at all the data as a block of numbers, we can calculate the central tendency of the dataset to help us describe it more easily.

The most common measure of central tendency is the **mean**, also known as **arithmetic mean** or **average**. This is calculated by adding up all the scores in the sample and dividing it by the number of scores. It is usually considered to be the best measure of central tendency, but there are some situations in which either median or mode is preferred. Mean has a companion statistic, standard deviation, which is used to show how far things vary from the average (see section 4.2.2.1).

Median is the midpoint or number in a distribution which has half of the scores above it and half below it if you sort the dataset in numerical order. Median is the preferred measure when there are a few extreme **outliers**, if there are some missing or undetermined values in your dataset, or if there is an **open-ended distribution**. An example of an **open-ended distribution** is if you were counting teacups in tea sets for a condition survey. Using categories of 1, 2, 3, 4, 5, or '6 or more teacups' would make it impossible to calculate the mean. You could also use median if you were using ordinal data, so if we go back to the condition survey example, you might score the condition of the teacups from 1 to 4, ranging from unacceptable to good, the data which represents the categories is ordinal data. Exercise 4.1 gives another example of these measurements.

Finally, **mode** is the number which occurs most frequently in a distribution of scores or numbers. Mode is used for nominal or ordinal data. It would be nominal data if a number is assigned to a category which has no rank, for example we might also decide to categorise our tea sets by colour, assigning 1 to red, 2 to blue, 3 to green, and so on. There is no correct order or rank for the categorisation, unlike the ordinal data in the condition survey example, where there is a defined progression in the data categories.

4.2.2 Measures of variability

It is useful to understand the measure of variability of a sample, because it will help you explain the results and allows you to look at whether your data is suitable for hypothesis testing. For measures of variability, we are going to look at three of the available measures: standard deviation, skew, and kurtosis. All three explore how spread out the data points are in your dataset. Standard deviation shows the variability of your data; **skew** shows whether your data is evenly distributed (normal) or whether it is clustered at the top or bottom of the range; and **kurtosis** shows whether it forms a pointy-shaped curve or whether it is flattened. Remember, these measures are used to provide a description of the datasets, helping you to summarise them into a manageable form, not as a way of comparing datasets or as a form of experimental analysis. Skew and kurtosis perform one other important role, as many statistical tests rely on data being normally distributed, and we can use these tests to check that.

4.2.2.1 Standard deviation

Standard deviation (SD) is one of the most commonly used measures of variability and looks at how spread out the data is; the more spread out the data, the greater its standard deviation will be. Standard deviation alone does not show that two datasets are significantly different. While having standard deviations that do not overlap can be indicative of a difference between the data, it cannot tell you that there is a **significant difference** without suitable inferential tests, as the standard deviation of two datasets can overlap even when the data is significantly different. Figure 4.2 shows two samples of data: both have a lot of cases (or data points) that cluster around the mean and have a few cases

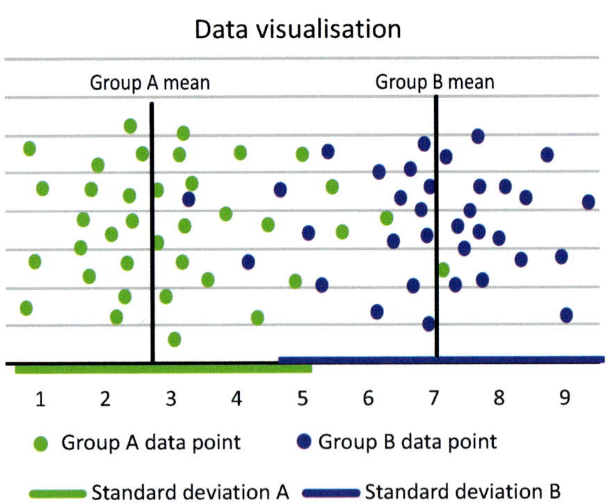

Figure 4.2 A representation of two datasets (green and blue spots). In both samples, the outlying cases result in a high and overlapping standard deviation (green and blue lines), but further statistical testing may show they are significantly different. Image by Philip J. A. Skipper.

which overlap with the measurements of the other sample. In a situation like this, calculating the standard deviation will give a large deviation within the sample, and if we plotted it on a bar chart, we would see the error bars from both samples overlapping (to be discussed in Chapter 5). However, these error bars do not reflect the majority of the data, and carrying out a hypothesis test would most likely show that the two groups were significantly different. When we want to present data graphically for the reader to compare between groups in a final output, we generally would not use standard deviation for the error bars, as it can lead to false assumptions; it is often better to use standard error instead (section 4.2.2).

Standard deviation is quite complicated to work out by hand, and the larger the dataset you have, the higher the chance of making a mistake, so it's best to use computer software to calculate it for you. To help you understand what is happening, it can be broken down into five steps as follows.

1. Calculate the mean for your dataset.
2. For every score in the dataset, find the square of its distance from the mean.
3. Add up all the values from step 2.
4. Divide the result from step 3 by the total number of scores in the dataset.
5. Calculate the square root of step 4.

We can compare our standard deviation to the mean, either visually, by plotting a graph with error bars, or by adding and subtracting from the mean, to see how variable the data is within the sample.

4.2.2.2 Standard error

Standard error of the mean (SE) is an inferential test, so it doesn't really belong in this section However, it is used, like standard deviation, to measure variability. It is also often used for error bars and as a descriptive measure, so it makes sense to introduce the concept here.

The difference between SE and standard deviation is that SE is used to estimate the variation of your sample mean when compared to the properties of the whole population, whereas standard deviation just looks at the variability within your sample. Because it is comparing your sample to the whole population, it evaluates how close your sample is to the whole population. As we don't know the mean for the whole population, we should properly refer to this as the estimated SE. This measure, unlike standard deviation, also gets more accurate the more samples you have and is used as a component of many hypothesis tests. To calculate the SE, we take the standard deviation and divide it by the square root of the total number of samples which were used to calculate the standard deviation. We will cover how to calculate this further in the supplementary material. You can see an illustration of the difference between standard error and standard deviation in box 4.1.

Box 4.1

During Philip's PhD, he measured the surface pH of damaged Lincoln limestone, comparing it to undamaged Lincoln limestone, to investigate whether microorganisms on damaged limestone were damaging the stone by secreting acids. The hypothesis was that if biological secretions were causing damage, the pH on damaged Lincoln limestone would be lower than on undamaged Lincoln limestone. The null hypothesis would be that there would be no difference in the surface pH for the two surface types. Eleven samples of each stone type were tested with each sample being measured four times.

Error bars based on the standard deviation of the samples (Figure 4.3, left) shows how much variability there is present within each sample, but you cannot use them to compare the two surfaces. Plotting the data using SE for the error bars (Figure 4.3, right) shows how the mean of the sample varies from the total population. Because SE is an inferential test, the reader can make a visual assessment about whether there is a genuine difference, which would then be further supported by the results of the hypothesis testing.

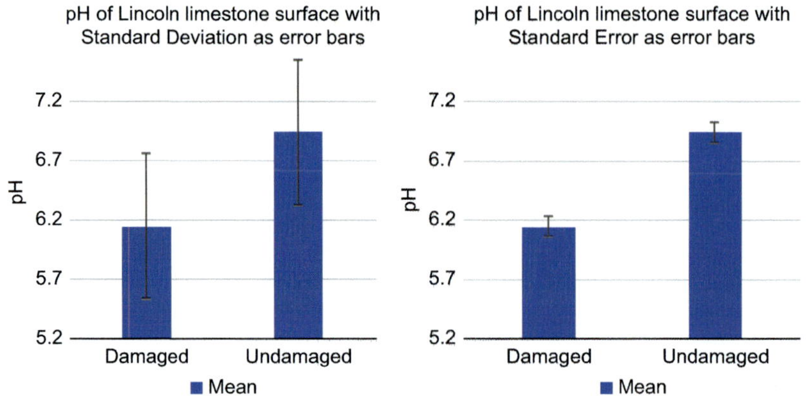

Figure 4.3 Left, pH of Lincoln limestone on damaged and undamaged stone, using standard deviation. Right, standard error bars compare the data to the inferred mean of the whole population. Image by Philip J. A. Skipper.

4.2.2.3 *Skew and kurtosis*

The more skewed your data is to one end or another when compared to the overall population, the less likely you are to get an accurate result from simple inferential tests such as the Student's *t*-test and ANOVA (see section 4.3). **Skew** can be

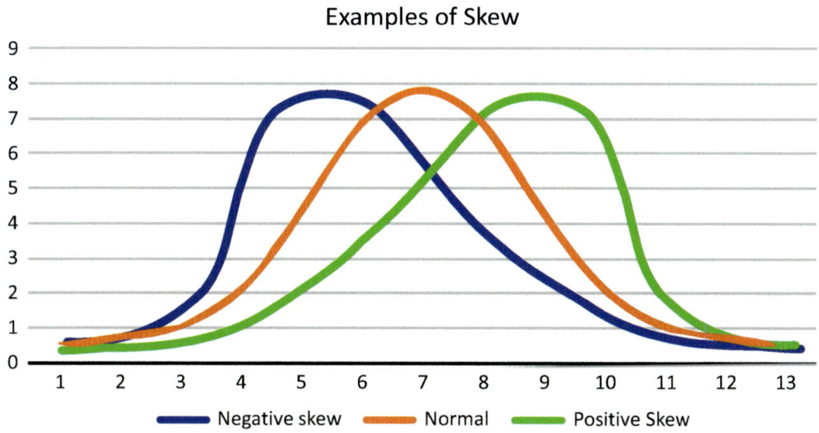

Figure 4.4 Examples of positive (green) and negative (blue) skew in data compared to a normal bell-shaped curve (orange). Image by Philip J. A. Skipper.

described as positive or negative and can be visualised on a graph (Figure 4.4) as having one steeply sloped side and one gently sloped side. If it is negative, then the steeply sloped side is closer to zero on the axis and vice versa. A **normal distribution** is evenly shaped and will produce a bell-shaped curve. An acceptable range for a skewness check is +2 to -2 (Hair *et al.*, 2010). If your data gives you a **skew** that is outside this range, then you need to talk to a data scientist about how to normalise your data.

We can also consider **kurtosis**, which is how pointy your data is. If it is abnormally pointy or abnormally flat, tests such as Student's *t*-test and ANOVA won't give you a meaningful answer. An acceptable range for these is +7 to -7 (Hair *et al.*, 2010). A 'pointier' dataset will result in a positive kurtosis result, and flatter data will give a negative kurtosis result. See Figure 4.5.

Because skew and kurtosis are used to describe the data, they are not tests which can be used to confirm that there is a significant difference between populations. However, we can use them as a basis to help our understanding and interpretation of the inferential tests.

If we revisit the data in box 4.1, calculating the skew allows us to interpret the different results. The damaged pH samples were negatively skewed towards the lower pH, and the undamaged samples were positively skewed to the higher pH (though they still fall in the +/-2 range for acceptable skew); where they cross over are outliers which result in a wider standard deviation. When we carried out inferential testing, the probability of the difference in pH occurring by chance, calculated with a Student's *t*-test (see section 4.3.1), was actually 1 in 136,131,000,000, or a *p*-value of 1.36×10^{-11} (to two decimal places), not something you would have predicted from looking at the standard deviation!

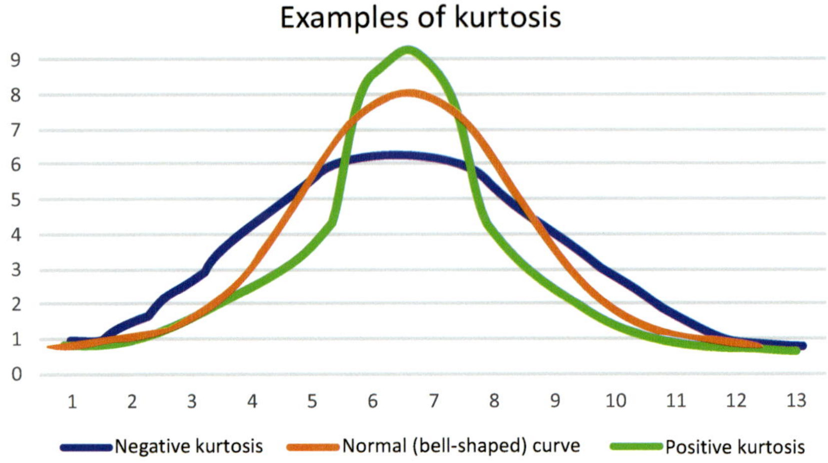

Figure 4.5 Examples of positive (green) and negative (blue) kurtosis together with a normal curve (orange). Image by Philip J. A. Skipper.

4.3 Hypothesis tests and statistics

When we test our hypothesis using statistics, we are measuring the probability that these results will apply to the whole population rather than just the sample and were not down to chance. In section 4.1.1, we defined *p*-value, which is the output that we will get from the tests discussed in this section.

There are three common hypothesis tests which we will look at: Student's *t*-test, ANOVA (both of which are for quantitative data), and Chi-square (which is for qualitative data and is pronounced 'ky', to rhyme with 'sky'). The results from these tests tell us whether we should accept our hypothesis or the null hypothesis (remember these from Chapter 1?).

Before we move on to the hypothesis tests, there is one last thing to define, and that is **degrees of freedom (*df*)**. When running statistical tests, we need to understand how much variability the test can cope with (i.e. the number of values which can vary) for the final calculation of the hypothesis test to still give an accurate result. Larger samples have larger degrees of freedom and (providing there is a genuine difference) will be more likely to generate a significant difference and prove the hypothesis. Degrees of freedom is a combination of how many cases you have in your sample and how many parameters in the population you are examining with your test. For example, if we were analysing three cleaning products in one test, we would have three parameters. Degrees of freedom is usually worked out by taking the number of cases in the sample and subtracting the number of parameters, so if we had taken 90 measurements of those same three different cleaning products, the degrees of freedom would be 90 – 3, or 87.

4.3.1 Student's* t*-test (for interval and ratio data)

> *'In vino, veritas'.*
>
> Pliny the Elder, 77 CE

At the start of this chapter, we quoted Andrew Lang, who cautioned us not to use statistics like a drunken man uses a lamp-post, but as Pliny the Elder said, there is truth in wine. Although to be completely accurate, the 'truth' we are about to look at originated in stout. In 1908, William Gosset, who was working for the Guinness brewery in Dublin, Ireland, developed a statistical test to help infer the quality of large batches of barley based on testing a small **sample**. While the Guinness brewery was happy for him to publish his work, they didn't want to give away the secrets of their quality control system, so he had to publish under the pseudonym of 'Student', hence the name Student's *t*-test.

Student's *t*-test requires the data to be **normally distributed**, which we would have checked first by testing the data with skew and kurtosis tests. Next, we need to decide whether the data is sampled independently or fully paired, and whether the variance (the calculated skew and kurtosis) in our data is equal or not. We discussed skew, kurtosis, and variance testing in the measures of variability section, so let's have a look at the other two requirements.

4.3.1.1 Paired versus independent data?

A fully paired test is one in which the test is carried out on the same subject before and after treatment. For example, if we are looking at surface hardness changes following the application of a new consolidant, then we would have two datasets:

1. Hardness of the samples before the application of the consolidant
2. Hardness of the same samples with the consolidant applied

Crucially, these are the same samples before and after testing, which is the key factor in determining that the data is paired.

What if we are concerned that the surface hardness testing would damage the surface and alter the post-consolidation results? In that case, we could use independent sampling, which would consist of:

1. Hardness of one set of samples which were tested without consolidant
2. Hardness of one different set of samples which were tested after consolidation

Ideally, both sample sets should contain the same quantity of samples, as this will give more robust results. This is not a problem with a fully paired test, but it's something which should be pre-planned if the experiment requires an independent test.

4.3.1.2 Carrying out a Student's t-test

Keeping with the example of the consolidant, how would we carry out our experiment and analyse the data? Because this experiment is going to be tested using a fully paired **Student's *t*-test**, we would take a reading from each **sample** before and after consolidation. Table 4.1 shows how the initial data might be recorded.

Because the measurements were carried out on the same **sample** material, a **paired Student's *t*-test** is the appropriate choice. A Student's *t*-test determines whether the means of each sample are equal (the null hypothesis) or whether they are not equal (the hypothesis). If the means are equal, it shows there is no difference between hardness before and after consolidation. Where they are unequal, it demonstrates that consolidation led to a difference in hardness. Table 4.2 shows the results after running a calculation on the numbers shown in Table 4.1.

The most important part of Table 4.2 is the ***p*-value**. Here, we have been given two ***p*-values**, one for a **one-tail** test and one for a **two-tail** test. If we think about the curves plotted on the graph in Figure 4.4, the tails are the left and right areas (starting from where the curve starts), or more correctly, the first and final 2.5% of the curve. In scientific research, the two-tailed test is used as standard because it tests the whole curve, so we will ignore the one-tail result, which will only look at either the left or the right of the curve. Another way of thinking about it is that a one-tail test will explore whether the mean is larger than the comparison sample OR whether it is smaller, but it does not test both. A two-tail test will look at whether the means are different – it simultaneously tests whether it is either larger or smaller.

When we look at the **two-tail *p*-value** in Table 4.2, we can see that it is less than 0.05, so there is a significant difference between the means of the two samples. If we look at the calculated means in Table 4.2, we see that without consolidant the hardness is 408.7, but after consolidation, it is 470.2. This shows the consolidant has caused a significant increase in the hardness of the surface, and we can accept our hypothesis.

Table 4.1 Sample of hardness data in Leeb hardness units from a consolidation study.

Sample number	*Before consolidation*	*After consolidation*
1	420	445
2	373	443
3	406	448
4	423	501
5	437	463
6	407	524
7	417	438
8	398	476
9	387	444
10	419	520

Table 4.2 Results from a paired Student's *t*-test generated in Excel using the analysis tools plug-in. The *p*-values have been highlighted in grey.[1]

Student's t-test: Paired Two Sample for Means	*Before consolidation*	*After consolidation*
Mean	408.7	470.2
Variance	353.1222222	1111.066667
Observations	10	10
Pearson Correlation	0.291555366	
Hypothesized Mean Difference	0	
df	9	
t Stat	-5.866610387	
P(T <= t) one-tail	0.000119385	
t Critical one-tail	1.833112933	
P(T <= t) two-tail	0.000238769	
t Critical two-tail	2.262157163	

The equations for the Student's *t*-test are quite complicated to follow manually, which is why we recommend using statistical software or an online calculator to carry them out. You can find a brief overview of the steps in the supplementary material in case you need to manually calculate it.

4.3.2 ANOVA (for interval and ratio data)

With the Student's *t*-test, we were comparing before-and-after results of one change. What if our test gives us multiple results per sample, or we have more than two samples? In this case, we would use an **ANOVA** (**analysis of variance**) test. There are two forms of ANOVA, a **one-way ANOVA**, which is the equivalent of a Student's *t*-test but for multiple samples, and a **two-way ANOVA** for testing multiple **samples** treated in different ways. ANOVA is a long test to work out by hand, so again, you can find more specifics about calculating it in the supplementary material. Here, we will look at the two types of ANOVA with an example of each to help you understand when to use them.

4.3.2.1 One-way (or single factor) ANOVA

Imagine an experiment in which you are comparing three different cleaning methods and taking surface roughness readings to look at whether they are damaging the surface of the material. The three cleaning methods are EDTA in water (this is a chelating agent, ethylene diamine tetra-acetic acid, often used in cleaning for removing surface soil from objects), saliva, and distilled water, all applied with a swab. You would also measure the surface before cleaning (this is the **untreated control**). Table 4.3 shows how the data might be recorded.

Table 4.3 Sample of surface roughness data for ANOVA testing. The full table is available in the supplementary materials for this chapter.

Untreated	*EDTA*	*Saliva*	*Distilled water*
0.86	0.77	0.83	0.81
0.87	0.86	0.82	0.85
0.86	0.75	0.90	0.84
0.89	0.90	0.91	0.81
0.85	0.86	0.92	0.84
0.86	0.73	0.91	0.93
etc.	etc.	etc.	etc.

Table 4.4 Results of single-factor ANOVA analysis generated by Microsoft Excel.

ANOVA: single factor						
Summary						
Groups	*Count*	*Sum*	*Average*	*Variance*		
Untreated	32	27.94	0.873125	0.000275		
EDTA	32	25.969	0.811531	0.005674		
Saliva	32	27.805	0.868906	0.001499		
Distilled water	32	27.726	0.866438	0.002056		
ANOVA						
Source of Variation	*SS*	*df*	*MS*	*F*	*p-value*	*F crit*
Between groups	0.081352	3	0.027117	11.41231	1.16 x 10–06	2.6776
Within groups	0.294642	124	0.002376			
Total	0.375994	127				

As the only difference between the samples is the cleaning agent, then the one-way ANOVA is the correct test to carry out. A one-way ANOVA tests whether the means of each group are all equal (the null hypothesis) or whether they are different (the hypothesis). Therefore, our null hypothesis would state there is no difference in mean readings of surface roughness for each cleaning product, while our hypothesis would be that there is a measurable difference between the mean readings for each cleaning product. Table 4.4 shows the results after running calculations with an extended version of the numbers shown in Table 4.3.

So what does all this mean? The summary table gives us a description of our data. For each sample set, it gives the total number of cases in the sample (count), the total of all the cases added together (sum), the mean (average) of the sample, and the variance of the values in each group. In our example, we can see that the EDTA group had the lowest sum and mean of all the groups and that the variance in all the groups was very low. Untreated had the highest sum and average, whereas saliva and distilled water were very close in the middle of the range. In general, the surfaces appear to become less rough after cleaning, with EDTA causing the most smoothing. The next table is used to tell us whether the pattern we have seen in the data is 'real' or significant, based on the results of the statistical test.

The ANOVA table is used to see whether there is a significant difference between the groups. We need to look at the *p*-value column, which is 1.16×10^{-6}, or 0.00000116, showing that there is a significant difference between the groups. The problem with interpreting this *p*-value is that although it says there is a difference, it doesn't tell us where that difference is. When we looked at the summary table, we could see that for the EDTA treatment, the **mean** roughness was lower than the other groups, so this indicates that there is a **significant** difference with EDTA cleaning when compared to saliva or water cleaning.

In our example, the difference between the groups was easy to spot, as the EDTA readings were lower than the mean roughness of other cleaning methods, but what if you wanted to check whether there was a difference between the cleaning methods and the untreated control, or if it wasn't so obvious? This is when we use Tukey's honestly significant difference *post-hoc* test (unsurprisingly, given that long title, it's generally referred to as Tukey's HSD). This also uses the data in some of the other columns on the ANOVA table (see also box 4.2).

4.3.2.2 Tukey's honestly significant difference post-hoc *test (Tukey's HSD)*

Tukey's HSD is used to find out the specific pairs of groups that generate a **significant** difference in an **ANOVA** test. All the information needed to calculate **Tukey's HSD** comes from the **ANOVA** test output tables.

The first thing to do is to set up the groups in pairs, as we need to carry out this statistical test for each pair. In this example, our pairs would be untreated and EDTA, untreated and saliva, untreated and water, EDTA and saliva, EDTA and water, and finally saliva and water.

Let's work through one of the pairs as an example. The first thing to calculate is the difference between the averages of the groups. Our first pair was untreated and EDTA. Using their averages from the summary table of the **ANOVA** (Table 4.4) gives a difference of 0.06159 (for this test, the difference needs to be positive, so always subtract the smaller number from the larger). We also need to calculate a **standard error** for the two groups, which is done by taking the 'within groups MS' figure from the **ANOVA** table and dividing it by the **sample** size for each group (which is listed as count in the summary table) then calculating the square root of the answer. For untreated and EDTA, this would be the square root of 32/0.00238, which is 0.0086. We then calculate the Tukey q score by dividing the difference (0.06159) by the error (0.0086), which, in our case, gives us 7.14.

Finally, we need to find the degrees of freedom (df) from the ANOVA table for the 'within groups' row, which is 124. If we look this up in a q table for a Tukey test, then for a df of 120 (this is the closest number on the Tukey test table to 124) and 6 samples (the number of pairs we were looking at), anything over 4.87 is a significant result to $p = 0.01$, so it shows there is a significant difference between the untreated and the EDTA cleaning methods. If we carried out this for all the pairs, we would see that it is only when we compare EDTA with one of the other cleaning methods that there is a significant difference in the surface roughness.[2]

Box 4.2 ANOVA table abbreviations

- *SS stands for sum of squares, that is the variation between the group means created by the levels of the independent variable and the overall mean.*
- *df is the degrees of freedom.*
- *MS is the mean sum of the squares and is calculated by dividing the sum of squares by the degrees of freedom.*
- *F is the overall F-value, which is calculated by dividing the MS for between groups by the MS for within groups and is used to look up the p-value if you were calculating by hand. You won't normally need this.*
- *F-crit is the critical value for F above which the p-value will be less than 0.05, so statistically significant.*

4.3.2.3 Two-way (or two-factor) ANOVA

With a one-way ANOVA, we were looking at one variable (in our example, different cleaning agents). A **two-way ANOVA** looks at two variables and compares them. It is possible to go beyond two variables, but that level of complexity is outside the scope of this chapter.

Building on our example for the **one-way ANOVA**, if we had used a swab or a sponge for cleaning different samples, we would have introduced a second variable. We would have two sets of surface roughness measurements (Table 4.5). These **sample** sets would be EDTA, saliva, and water cleaning carried out with a swab; EDTA, saliva, and water cleaning carried out with a sponge, and of course the untreated **controls**. In this case, swab versus sponge would be our first variable and the cleaning agents our second variable.

Table 4.5 Sample of data for two-way ANOVA testing. The full table is available in the supplementary materials for this chapter.

Swab				*Sponge*			
Untreated	*EDTA*	*Saliva*	*Distilled water*	*Untreated*	*EDTA*	*Saliva*	*Distilled water*
0.86	0.77	0.832	0.808	0.86	0.735	0.909	0.86
0.87	0.857	0.817	0.849	0.88	0.8	0.937	0.823
0.86	0.748	0.898	0.84	0.88	0.726	0.938	0.831
0.89	0.904	0.909	0.813	0.90	0.906	0.894	0.829
0.85	0.856	0.917	0.836	0.87	0.725	0.934	0.879
0.86	0.733	0.912	0.932	0.85	0.92	0.943	0.801
etc.	etc.	etc.	etc.	etc.	etc.	etc.	etc.

When we looked at the one-way ANOVA, we defined the hypothesis and null hypothesis. With a two-way ANOVA, it gets a little bit more complicated, as there are three possible hypotheses and null hypotheses.

The hypotheses are as follows:

1 There is a difference in the means of the first variable (e.g. there is a difference depending on whether sponge or swab was used).
2 There is a difference in the means of the second variable (e.g. there is a difference in surface roughness depending on the cleaning agents).
3 The effect of the first variable is dependent on the effect of the second variable (e.g. using a sponge makes the EDTA more effective than using a swab).

The null hypotheses are as follows:

1 There is no difference in the means of the first variable (in our example, the different application methods, swab or sponge).
2 There is no difference in the means of the second variable (in our example, the different cleaning agents).
3 The effect of the first variable (sponge versus swab method) does not depend on the effect of the other variable (cleaning agents). In our example, this would mean there is no difference between using the swab or the sponge for any of the cleaning agents.

Since we have the two sets of samples, swab and sponge, both of which contain the same cleaning agents, we would carry out a two-way ANOVA. When using software, our results would be generated in a set of tables similar to Table 4.6.

Just like the one-way ANOVA, we have summary tables and ANOVA tables. For a two-way ANOVA, we have three summary tables, one each for the first variables (swab and sponge). Each of these contains the second variable information (untreated, EDTA, saliva, distilled water, and a total column for all four). We also have a total table, which summarises the first two tables.

If we look at the final ANOVA table, we now have four rows, of which the first three are testing the three hypotheses. 'Sample' generates a *p*-value for testing hypothesis 1, 'columns' gives the *p*-value for testing hypothesis 2, and 'interaction' shows the *p*-value for hypothesis 3. The other components of the table are the same as for the one-way ANOVA. Again, the *p*-value is the column of most interest to us; as it is greater than 0.05, it tells us there is no difference between the samples (sponge versus swab). There is a significant difference between the columns (cleaning agents), but it doesn't tell us which one is significant, so we would use Tukey's HSD to calculate this. Finally, we have the interaction *p*-value, which explores whether there is a significant link between the samples (sponge *vs* swab) and the columns (cleaning agents); in this case, there wasn't one, as the *p*-value is greater than 0.05. However, if for example EDTA was more effective with a swab and saliva more effective with a sponge, this would be reflected in a significant *p*-value here, and Tukey's HSD would be used to work out which combination

Table 4.6 ANOVA calculated tables for the dataset from Table 4.5.

ANOVA: two-factor with replication						
Summary						
Swab	*Untreated*	*EDTA*	*Saliva*	*Distilled water*	*Total*	
Count	15	15	15	15	60	
Sum	13.067	12.104	13.099	13.007	51.277	
Average	0.871133	0.806933	0.873267	0.867133	0.854617	
Variance	0.000321	0.005329	0.001368	0.002512	0.003037	
Sponge						
Count	15	15	15	15	60	
Sum	13.133	11.924	13.406	12.826	51.289	
Average	0.875533	0.794933	0.893733	0.855067	0.854817	
Variance	0.000239	0.004625	0.002201	0.000984	0.003316	
Total						
Count	30	30	30	30		
Sum	26.2	24.028	26.505	25.833		
Average	0.873333	0.800933	0.8835	0.8611		
Variance	0.000276	0.004842	0.001832	0.001725		
ANOVA						
Source of Variation	*SS*	*df*	*MS*	*F*	*p-value*	*F crit*
Sample	1.2E-06	1	1.2E-06	0.000546	0.981397	3.925834
Columns	0.123254	3	0.041085	18.69757	6.67E-10	2.685643
Interaction	0.005458	3	0.001819	0.827928	0.481184	2.685643
Within	0.2461	112	0.002197			
Total	0.374812	119				

was the significant one. The full calculation for all this data is available in the supplementary material.

4.3.3 X^2 tests (Chi-square tests, for nominal and ordinal data)

When we were looking at measures of central tendency in section 4.2.1. we talked about ordinal and nominal data, which is where you apply a numerical value to a category. To test a hypothesis where the study has used ordinal or nominal data, you use the **Chi-square** test. It is important to note that this test must be carried out on the actual number of occurrences (observed counts): you can't use it on percentages, proportions, means, or anything else. If you wanted to make comparisons between the number of items containing copper in a collection which was measured using chemical spot tests where the results were either positive or negative, then the Chi-square test would be the appropriate one for those results.

The Chi-square test does not work well with small numbers. This is where planning the analysis alongside planning the experimental research is vital, because you need to design your experiment to ensure that you will have greater than five observed counts per category.

In section 4.2.1, we talked about categorising tea sets by colour, so let's assume you are presenting a paper on the change in colour preference in tea sets between the 18th and 19th centuries. Your hypothesis is that there was a significant change in preferred colour between the 18th and 19th centuries, whereas the null hypothesis would be that there was no change. You decide to back up your observations with some statistics. Because the results are ordinal data, you would use a Chi-square test. The first step is to place the counts into a table (as shown in Table 4.7) and work out the total numbers for the rows and columns. These are your observed counts.

Next, you need to work out the expected frequencies for each category, i.e. 18th century + Red or 19th century + Brown, and place them in a second table using the following equation.

$$\frac{\textit{Total for row (18th or 19th C)}}{\textit{Total (1730 in this case)}} \times \frac{\textit{Total for column}\left(\textit{Red, green, e.t.c.}\right)}{\textit{Total (1730 in this case)}} \times \textit{Total (1730 in this case)}$$

To work out the expected frequency for 18th century Red, you would use:

$$\frac{950}{1730} \times \frac{520}{1730} \times 1730 = 285.549$$

A full set of expected frequencies is shown in Table 4.8.

For each category, you then subtract the expected frequency from the observed count, square the result, and divide it by the expected result. This gives us the difference between the measured frequencies and the expected frequencies.

$$\frac{\left(\textit{observed count} - \textit{expected frequency}\right)^2}{\textit{expected frequency}}$$

Table 4.7 Tea sets in a collection categorised by colour and date.

	Red	*Green*	*Blue*	*Brown*	*Total*
18th C	500	200	200	50	**950**
19th C	20	500	250	10	**780**
Total	**520**	**700**	**450**	**60**	**1730**

Table 4.8 Expected frequencies of tea set colour and date, calculated using the data from Table 4.7.

	Red	*Green*	*Blue*	*Brown*	
18th C	285.5491	384.3931	247.1098	32.94798	**950**
19th C	285.5491	384.3931	247.1098	32.94798	**780**
	520	**700**	**450**	**60**	**1730**

Table 4.9 Difference between expected and actual tea set colour and date frequencies.

	Red	*Green*	*Blue*	*Brown*
18th C	161.0552	88.45321	8.981172	8.82517
19th C	246.9499	34.769	0.033803	15.98306

Table 4.10 Extract from Chi-square table, showing degrees of freedom on the left column and *p*-value in the top row. The degrees of freedom equal to those used in our example (row 3) are highlighted in grey. A full version of the table is available in the supplementary material.

df	*0.995*	*0.99*	*0.975*	*0.95*	*0.90*	*0.10*	*0.05*	*0.025*	*0.01*	*0.005*
1	—	—	0.001	0.004	0.016	2.706	3.841	5.024	6.635	7.879
2	0.010	0.020	0.051	0.103	0.211	4.605	5.991	7.378	9.210	10.597
3	**0.072**	**0.115**	**0.216**	**0.352**	**0.584**	**6.251**	**7.815**	**9.348**	**11.345**	**12.838**
4	0.207	0.297	0.484	0.711	1.064	7.779	9.488	11.143	13.277	14.860
5	0.412	0.554	0.831	1.145	1.610	9.236	11.070	12.833	15.086	16.750

So if we do this for 18th century Red, we would use

$$\frac{(500-285.5491)^2}{285.5491}=161.0552$$

Repeating this for all the combinations gives us the following data, shown in Table 4.9.

We then add all the numbers in Table 4.8 together to get our Chi-square statistic, which is 565 in this case. This is not the *p*-value; to find this, we need to work out the degrees of freedom and then look up the *p*-value on a Chi-square test table.

For the Chi-square test, we use a different formula to calculate the degrees of freedom, which is calculated by the following formula:

$$df=(rows\,in\,table-1)\times(columns\,in\,table-1)$$

In Table 4.7, we have four columns and two rows (we ignore the 'total' rows and columns here), which gives us 3 degrees of freedom (4 – 1 multiplied by 2 – 1 = 3 × 1). Going down the left-hand column of the Chi-square table (Table 4.10), we have degrees of freedom. In our case, we need the row labelled 3. The columns contain the Chi-square result for different *p*-values, so we read from left to right along df row 3 until we find one which is lower than our result. In this case, our result is higher than all the Chi-square results, so we say that the *p*-value is less than 0.005 (shortened to $p < 0.005$). This shows that we can reject the null hypothesis and that there is a significant change in the colours of tea sets between the 18th and 19th centuries.[3]

This test can also be carried out using a statistics package, which will generate a *p*-value without the need to look it up on a table.

4.5 Summary

This chapter was designed to give an overview of which tests are appropriate for different types of data and how to interpret the results from them based on an initial hypothesis. We have not gone into depth here about how to carry out all the tests, because otherwise, this would be a book entirely about statistics, with no room for anything else! The main guides for carrying out the tests are provided as supplementary materials on the accompanying website, so please do look through them.

In the chapter, we have discussed why you should use statistics and the difference between descriptive and inferential statistics and looked at some briefly worked examples to help explain them. With time and patience, you can carry out any of these tests by hand, but it is much easier to use a spreadsheet or statistical package to get the results.

Most spreadsheets have built-in formulae for these tests, but they won't always give you all the information, because for some tests, the only output will be the *p*-value. There are free statistical packages such as R, or it might be that you work at an institution which gives you access to packages such as MATLAB and SPSS. The most important aspect is to select a package that you are comfortable with and can get good support for. These computer packages are only as good as the data that is put in, so it is important to make sure that everything is correctly set up with your experiments to generate good data which is accurately recorded. Some tests such as Tukey's HSD are also not built into most spreadsheet software, which is why we have covered that in the chapter and supplementary material in more detail.

In the supplementary material, we have provided exercises which work through each of the tests mentioned in the chapter using Microsoft Excel and Google Sheets and highlight any differences between the two.

Exercises

4.1 Mean and median calculations

Have a go at calculating the mean and median of this data, which are weight measurements of objects in a collection. How might the different types of information be useful?

Accession number	*Weight in kilograms*
2023.108	15.6
2023.109	20.02
2023.110	18.9
2023.111	110.5
2023.112	19.1

4.2 Will these examples of research projects generate qualitative or quantitative data or both?

- a How has public opinion influenced decisions to repatriate stolen or looted art?
- b What is the condition of a museum's collection?
- c How might we conserve Saxon grave goods excavated at Lincoln Castle?
- d How widespread is the 'volunteer conservator' in the UK heritage sector?
- e How important is 'conservation in action' in modern conservation practice?

4.3 Data types

Read through the table and work out the type of data that each section refers to. Are they ordinal, nominal, ratio, or interval data?

Type of data	*Definition*	*Example*
a	Labelled groups, which can be read in any order	Categorising ceramics by manufacturer or shape, or sculptures by stone type
b	Labelled groups with defined order of categories	Survey of condition of paintings, using condition grades
c	Interval data, which can be positive or negative	Temperature readings in degrees Celsius taken from different rooms in a building
d	Interval data, which can only be positive	Wood moisture readings taken over a period of time

4.4 Choosing hypothesis tests

Look back at your answers from Exercise 4.3. Which hypothesis tests that we have discussed in the chapter would be appropriate for the different data types?

Notes

1 Data is shown here and in other tables with all decimal places so that you can see the types of output you will get from running the tests. When presenting the final results, you would normally just show one or two decimal places.

2 This data is given as an example only and does not represent findings from the treatment of any object.

3 This example uses fictional data.

5 How to display results

Introduction

Why do we need to present data? Data is the information gained from your work, and the data needs to be presented in some format to help the audience (this term may include readers as well as those seeing a presentation) follow and understand how this has been used to reach conclusions. How people present their data can make the difference between the audience understanding the argument and being confused about the outcome of a study. When data is presented well, it helps to build confidence in the reliability and validity of the work. It may also make the difference between a paper being accepted for publication or rejected.

Firstly, though, we will consider what types of data might constitute results. We have already explored the concepts of qualitative and quantitative data and what techniques might be used to obtain that data. We've also looked at statistical analysis and the types of data generated from statistical testing. The key point for this chapter is that any information obtained from your own experimentation, whatever that technique happens to be, forms your result or a set of results.

It is key for the writer or presenter to understand who the audience is for this data. Well-presented data helps audience engagement with the message that the writer or presenter is aiming to communicate. However, presenting data for a public talk might look very different from the kind of data that you may present in a scientific journal. That journal article data may be very clear to a specialist audience and be exactly what's required for the publication, but it would leave a non-specialist audience with no experience in the field confused and disengaged. A report for a funder or a private client may be different again. We will further discuss the topic of audience and what that means for your writing style in Chapter 6.

This chapter considers a variety of ways in which data might be presented. It will not present all the options for every type of data, since they are far too varied to cover in a short chapter. Our aim is for the reader of this book to have some understanding of the different formats and when they might be used so they can then apply this to their own results.

DOI: 10.4324/9781003397137-8

5.1 Raw and processed data: terminology

When we consider data, it is important to make the distinction between raw and processed data. **Raw data** is the data that comes directly from the source. An example of this could be the report file or exported data from an XRF or FTIR analysis (such as shown in Figure 5.1) or a set of readings of surface roughness data for a stone surface. This data would not have been checked through for errors, for example errors in data collection causing an outlier, as considered in Chapter 3. Other terms sometimes used for raw data are 'primary data' and 'source data'.

There are times when the raw data forms the basis for the information presented to the audience; for example, a microscope image is primarily raw data and could be presented with minimal alterations in a report – perhaps with added annotations and a scale bar.

More commonly, though, the **processed data** is what's presented as the results from a study, though it may be appropriate to show the source of that processed data through including raw data in an appendix or as supplemental information in an online publication. Processed data has been checked through for any results that are excluded, and it may have been analysed in a range of different ways to get meaningful information from it and form conclusions. An example of this could

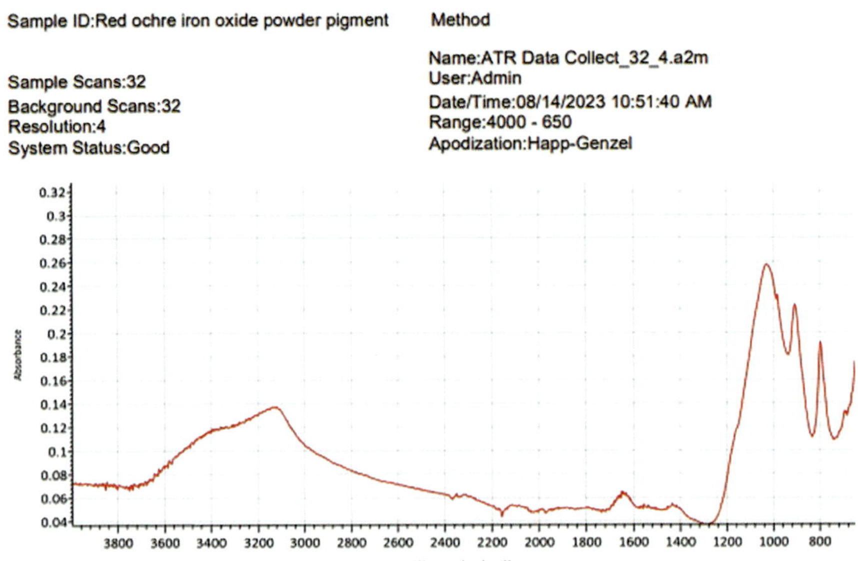

Peak Number	Wavenumber (cm^{-1})	Intensity

Figure 5.1 Raw data generated from FTIR analysis of red ochre, an iron oxide powder pigment, in the format provided by the manufacturer's software (Agilent). This data is unprocessed, as peaks have not been assigned, and also has an accompanying data file of all the wavenumber and absorbance readings. Image by Lynda K. Skipper.

be taking an average value for the stone roughness readings that we mentioned under raw data and comparing it statistically to the average roughness reading for a different type of stone. The actions taken to interpret this data are known as **data processing**, and the result of the processed data is sometimes described in the literature as 'information'.

5.2 What might poorly presented data look like?

In the introduction, we discussed how poorly presented data can be challenging for the audience to read and understand. Let's look at an example of this in practice using a simple survey (note that this data is fictional!).

This survey data looks at the types of cakes that conservators prefer. In a survey of UK conservators, they were asked to pick their favourite type of cake. Seventeen conservators said that they did not like cake at all. Thirty-four conservators picked the chocolate cake, 28 conservators picked a carrot cake, and 7 conservators went for banana cake. The fruit cake was fairly popular, with 16 conservators choosing that one, though it was less popular than the chocolate and carrot cakes, and 22 people liked Victoria sponge the best. Only 5 people said they liked coffee and walnut, so that was the least popular.

Although the text is designed to be fairly readable, compare this to a **pie chart** giving the same data, and take a moment to consider what you think about the two styles of presenting data in text compared to graphically (Figure 5.2). This pie chart

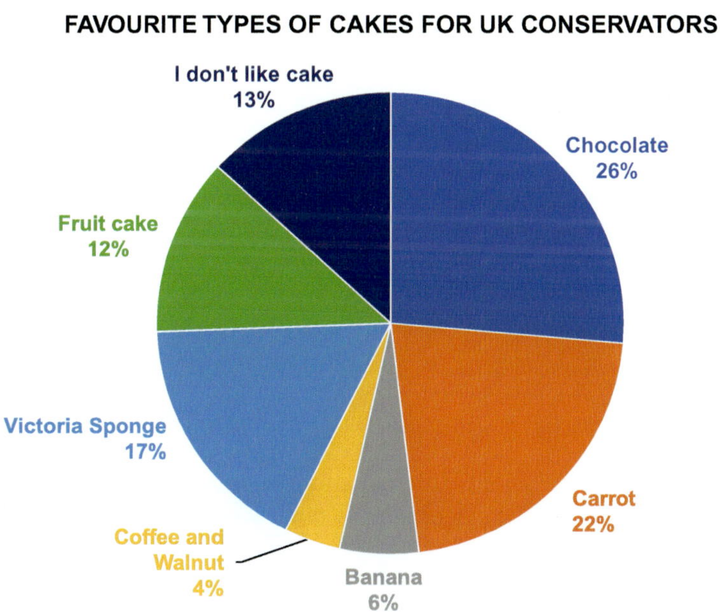

Figure 5.2 Favourite cakes chosen in a survey by UK conservators. Each section shows respective percentages of conservators who chose that type of cake, with chocolate being the favourite. Image by Lynda K. Skipper.

data has also been further processed to show it as percentages (though perhaps to fit with the graph type, it should have been a survey of favourite pies).

Viewing something visual with clearly laid-out information is generally a more accessible format than a block of text presenting a series of numbers. Although some interpretation was given in the text, it's easier and quicker for the reader to follow the argument and cross-compare the results when they are laid out visually. Figure 5.2 would still need some text highlighting key points, but it would be shorter because the reader does not need the detailed description of each value.

It's also important to filter the data. In general, studies generate a lot of information. Another example of poorly presented data would be information overload – trying to do far too much within the word count or space available. Figure 5.3 shows an example that Philip didn't use in his PhD, which was too information dense to be readable. The way you present data needs to be selected to ensure that it provides evidence to support or disprove the hypothesis and answers the research question of the study. This would usually be your processed data, but it need not be all of it. It's not unusual to find that not all the data you have collected is relevant. What you must not do is selectively choose to show only the data that supports your hypothesis and ignore the data that disproves it. That leads to flawed conclusions and means your work is not valid.

5.3 Use of tables within the results

Although the examples shown have demonstrated an example of a written format and a graph, it is helpful to consider also whether data might be successfully presented in a table. Tables are a way of showing data in an organised and more accessible fashion, perhaps through grouping or listing information. We find tables are particularly helpful for presenting numerical data that is not appropriate for a graph or summary data for a range of different tests. Graphs are helpful for a visual representation of data, but where it's important to see precise numbers, a table may be able to represent that more clearly.

Tables may contain numbers, text, or a combination of both. This is also a useful property of tables, since some data may have combined types of information, which makes it difficult to show solely in a graph, such as the example shown in Table 5.1.

5.4 Use of graphs within the results

Graphs are a way of presenting different information and collating it into a more visual format, again with the aim of it being more accessible and organised. As we have seen from the pie chart, they can allow for comparisons between different types of data within a dataset or between datasets to help draw conclusions about the findings of the experiments.

A pie chart shows the data in a circular format. It uses 'slices' of the pie to show a relative proportion of that value. For a fairly simple dataset, this style can

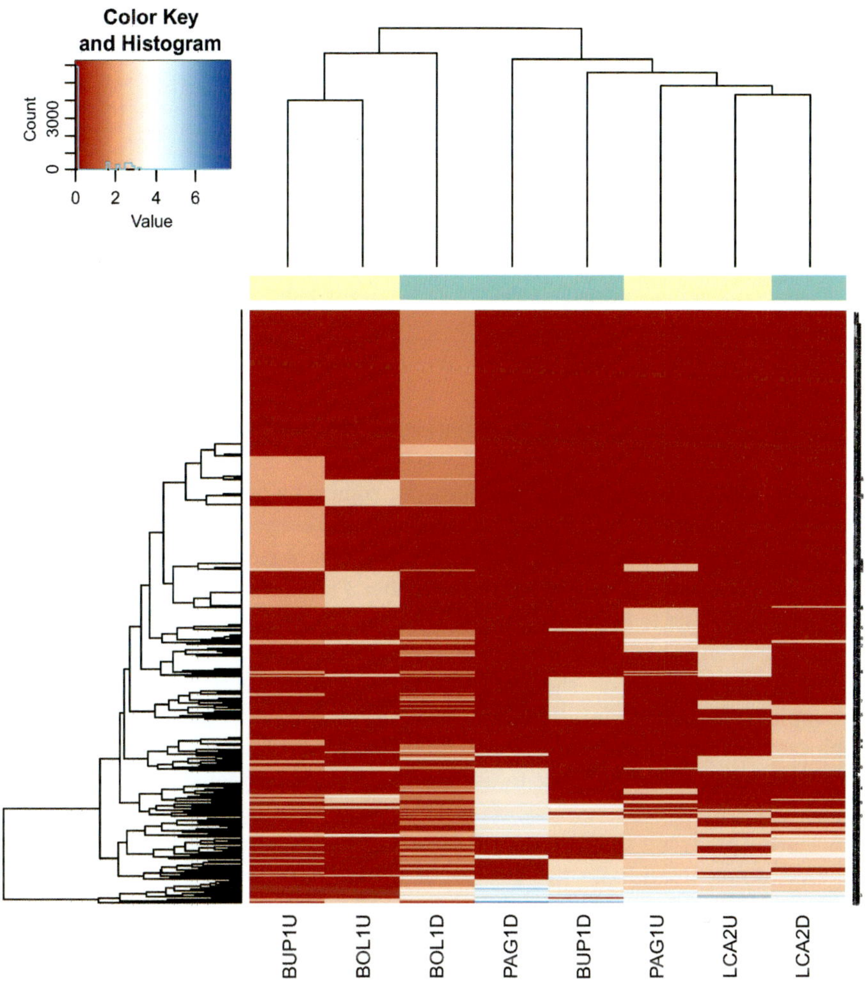

Figure 5.3 This is a heat map and distance tree showing which species of bacteria were found on different stone types at different locations. The illegible blur down the right-hand side is the names of over 1,000 different species of bacteria. In the end, Philip chose to remove the heat map and just use the tree at the top to show how the different sites were linked to each other and to present the data on the bacterial species in a different format. Image by Philip J. A. Skipper.

Table 5.1 Summary data from iron spot tests showing presentation of combined quantitative and qualitative data.

Item tested	*Spot test used*	*Colour of result*	*Overall result*	*Approximate quantity based on colour*
UL23/041	Iron	Blue	Positive	25 mg/L
UL22/016	Iron	White	Negative	N/A

be quite effective to show preferences or divisions between categories. Yet if we return to our example in section 5.2, what if we'd allowed our imaginary conservators in the survey to choose a first-preference cake and a second-preference cake? That would probably have suited Lynda better, since she is a fan of both chocolate and coffee-and-walnut cake . . . but that data, if placed in a pie chart, would not be easily comparable. The data could not all be added together into one single pie chart, as there are too many different categories, and they do not add up to 100%. Perhaps we could draw two pie charts, one with first-preference cake and one with second-preference cake, but even so, examining the data in detail would involve more work for the reader than necessary. It is also not always necessary to use a pie chart if there are only two categories, such as the equivalent of yes-or-no or positive/negative data; perhaps a table would show this in a more condensed and straightforward format. Equally, if there is a very large number of categories, it is important to be cautious about presentation, as many small categories can easily become 'swallowed up' in the visuals because of the insignificant size of the slices.

This is why it's so important to choose the right type of graph for the project at hand; every format of graph has advantages and disadvantages, and some are more suited to particular types of data than others. Pie charts are, of course, not the only option. In work that we have reviewed, bar charts and line or scatter graphs, along with pie charts, tend to be the main graph types used. These are probably the easiest to create in a range of graphical software, and this chapter will not describe how to use that software. However, there are a range of resources in both the exercises and the supplemental online information, which gives you the opportunity to practice graph creation. In this section, we will consider some of the different graph types and why someone might choose to use them and also consider some lesser-known alternatives.

A **scatter plot** plots the data at each measured point. A **line graph** can be used to join these points up. It looks like a series of points connected by a single line, often showing change of a variable over time, and uses a two-axis graph format, with axes (which is the Latin plural of 'axis') commonly referred to as x (the horizontal axis) and y (the vertical axis). There may be more than one line shown on a single graph. The image shown in Figure 5.1 is a line graph, but in this case, it is made up of hundreds of data points for absorbance, taken over different wavelengths. Figure 5.4 shows a range of features on a graph alongside the terminology you can use to describe them. Being able to use and understand this terminology will help you with interpreting what you see in the data. The **trend line** follows the general route of the data without necessarily going through all the data points.

Dual-axis charts extend this concept further by having two y axes, and conservators may well be familiar with this format from looking at graphs showing RH and temperature on a double y axis. They both use the same x axis measurement (time) but in this case, having two y axes allows different scales to be shown for the different variables and in a comparable format. An example of this is shown in Figure 5.5.

A **bar chart** is a series of bars of the same width, generally extending from the x axis, with the height on the y axis indicating their value. These bars represent

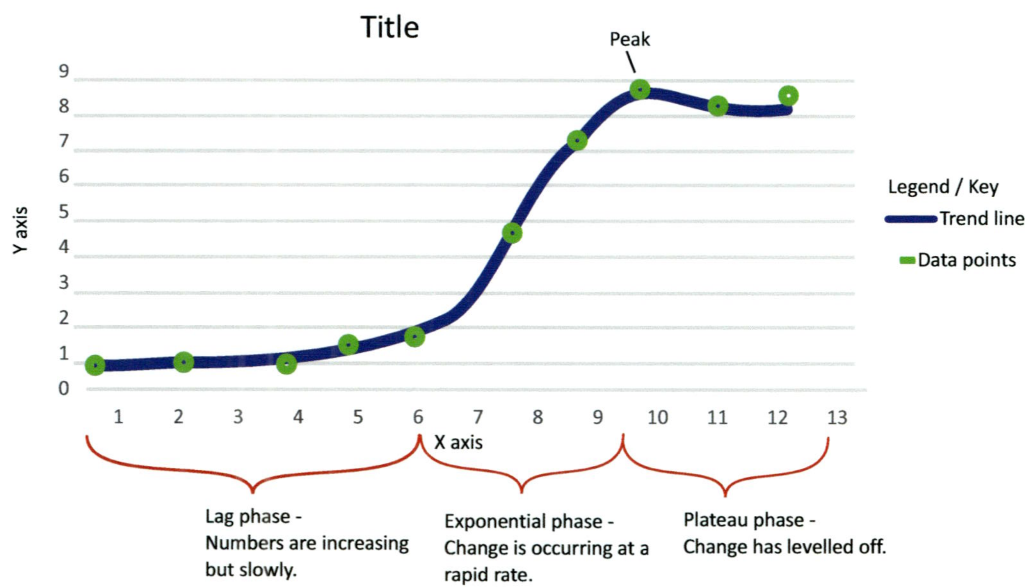

Figure 5.4 Reading a graph, illustrated by a line graph sketch showing labelled features. Image by Philip J. A. Skipper.

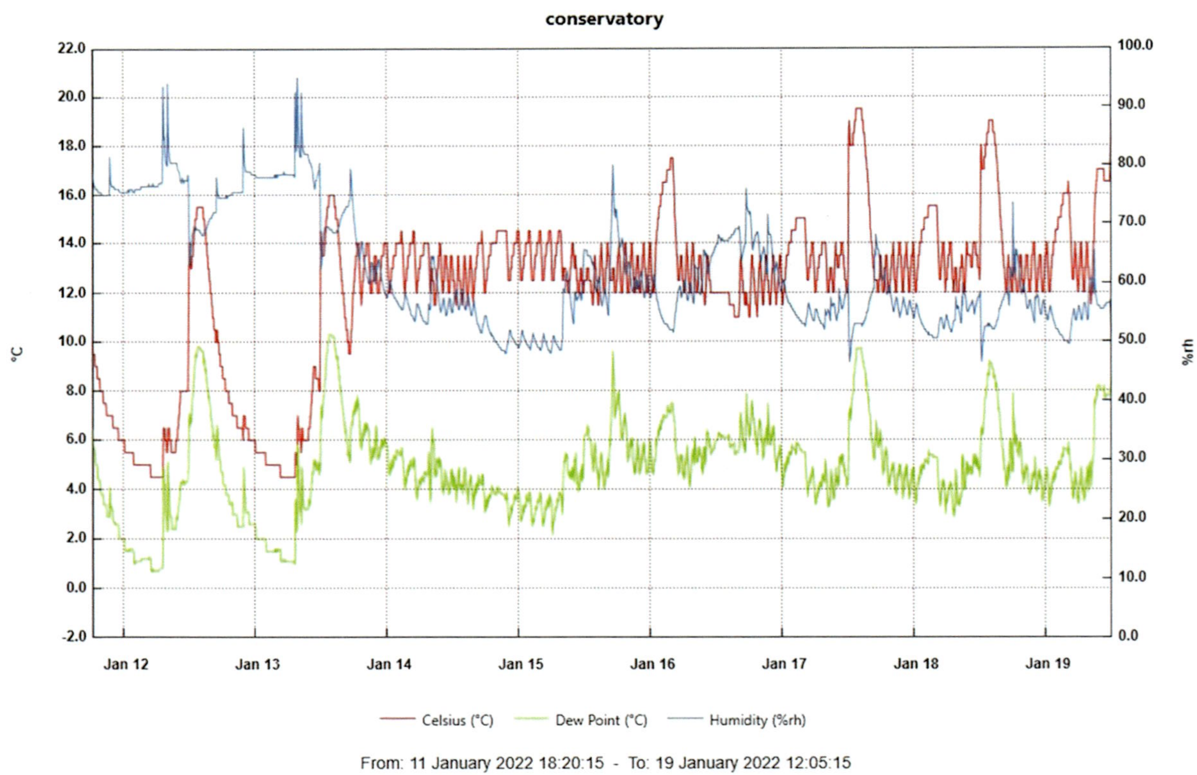

Figure 5.5 Raw data for RH and temperature on two different y axis scales, measured over time (x axis). This data is testing the effect of different radiator settings in our conservatory. Given the high RH levels and large fluctuations, conservators will probably be reassured to know that no historic objects are kept in there! Image by Lynda K. Skipper.

individual measured points or a series of measurements for each variable. Less commonly, a **vertical bar chart** can be used, where the bars extend from the y axis instead of the x axis. **Bar charts** are useful when the data is not linked and cannot be joined by a line. They can also be helpful for showing change in a measurement over time. The average change in surface colour after fading of a watercolour paint, for example, could be shown as a series of bars at different RH values.

Showing data graphically is not just useful in a scientific context. Lynda has recently been working through some archival business data as part of her research. Within the data, there were figures relating to company income for a range of different years for one of their products. If we take one of these years, 1929, we can use a bar chart to see how the income varied through the year (Figure 5.6). Potentially, extra bars could be added next to them for other years, plotting 1928 and 1930, for example, to see how income varied in different years. Because the data is linked, an alternative presentation could be to use a line graph with data points each month. However, the total income data is actually composed of three different income streams. On a standard bar chart, this could be shown as three separate bars.

A different way of presenting this data could be to use a **stacked bar chart**. This shows visually how that income is broken down into three different aspects (sales of product, packaging cost, and postage) each month (Figure 5.7). The type of graph you might use here would depend on the key message you wish to communicate, and this relates to having a clear argument or purpose to each figure illustrated with a graph. In this case, the key message in this chapter is that visual

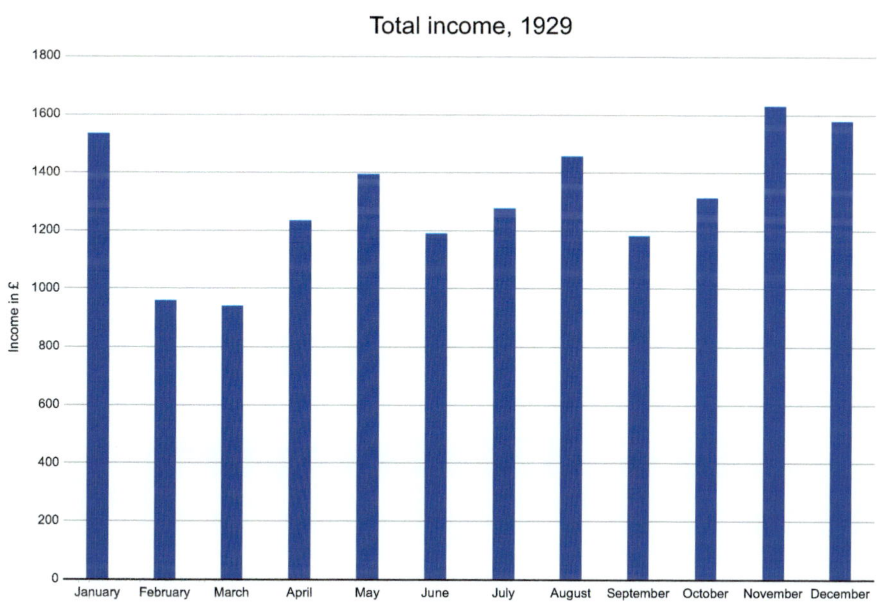

Figure 5.6 Bar chart showing income for 1929 financial year. Image by Lynda K. Skipper.

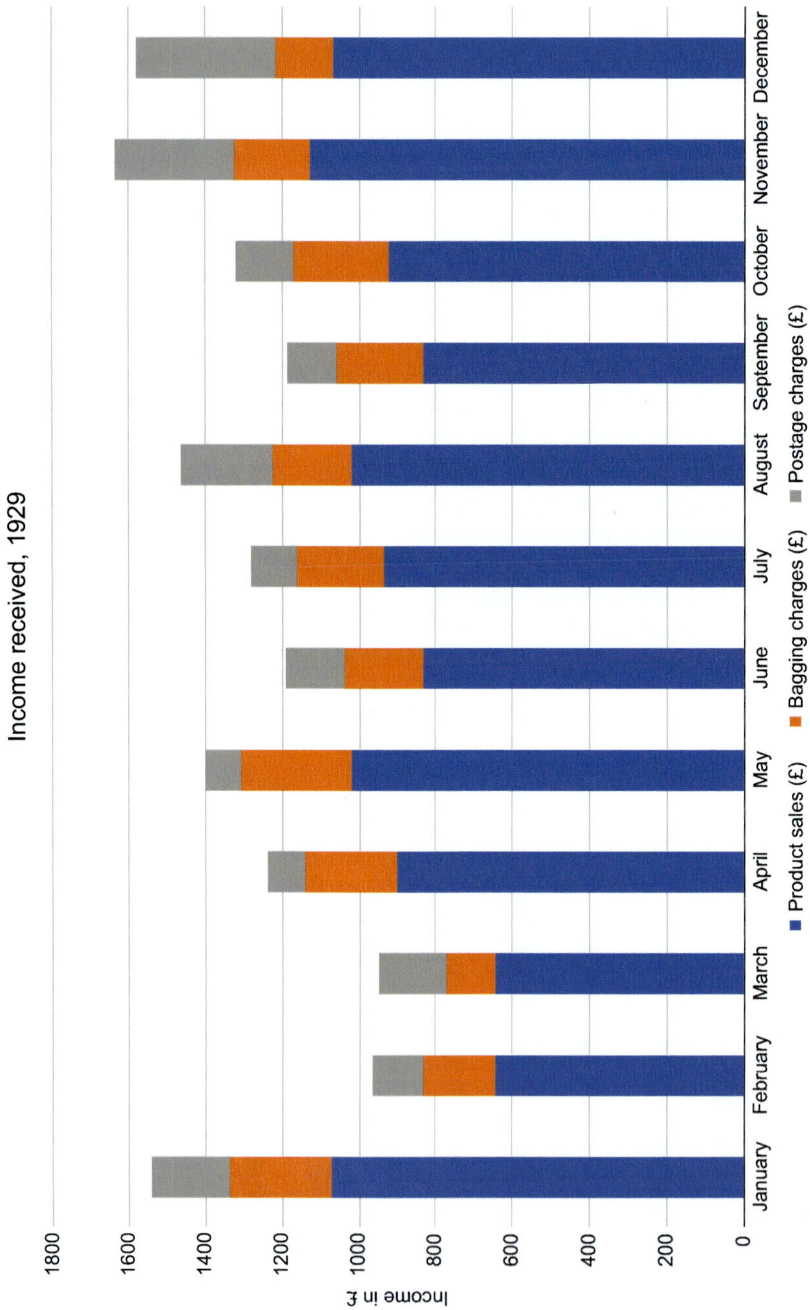

Figure 5.7 Stacked bar chart breaking income down into the different categories, which gives more detailed information about different income streams than the previous chart. Image by Lynda K. Skipper.

communication of data is great, but you need to think carefully about what message you wish to convey and pick a graph type accordingly!

The problem with bar charts comes when there are too many variables, because the graph can become overcrowded. If we take the same premise of fading watercolour paint, then if we were to measure fading at different RH levels, for five different colours of paint, while also measuring the fading over time with visible light, UV light, and in combination, you can perhaps start to imagine how chaotic it would look to squeeze all of that data into a single bar chart. It might also suggest someone was trying to cover too many different variables at once, and they should perhaps revisit their experimental design, but that is not the key point of this section.

With both bar charts and line graphs, it is possible to add **error bars** at the data points. It is important to apply the correct type of error bars. In Chapter 4, we discussed standard deviation and standard error. Both can be used as error bars, and which type to use depends on the message you are trying to convey to the reader. If you want the reader to make a comparison between quantities on a bar chart, then standard error would be the correct choice, since this is an interpretive statistic (Figure 5.8, left). If you are presenting XRF data in a graph and you want to show the variance for each element, then you would use standard deviation (Figure 5.8, right). This can be a very useful feature, which again helps establish confidence in the result. Where the wrong error bars are selected, however, the overlap can lead the reader to believe that the differences are unlikely to be significant – though of

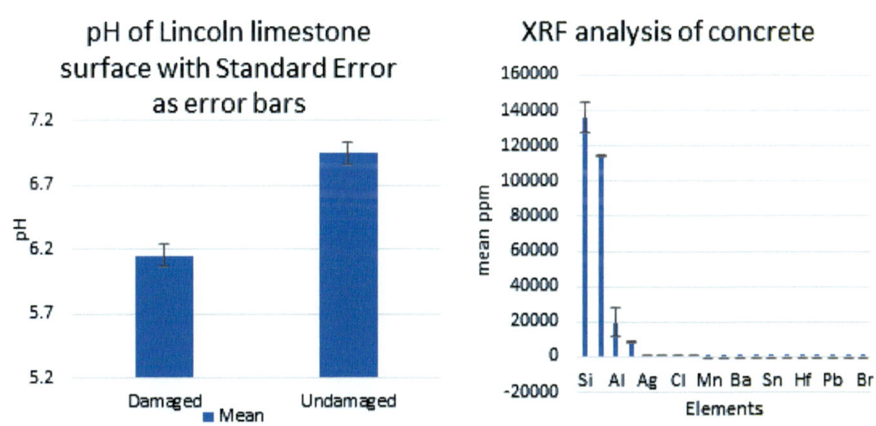

Figure 5.8 In the graph presenting the limestone pH, we want the reader to compare the two bars; as they are comparing between samples, then we use standard error for the error bars. With the XRF analysis, we have given the mean parts per million (ppm) for a group of samples, and we want to show how much variation there was within the samples for silicon or for other elements, so we use standard deviation; the expectation was never for the reader to try and see if two of the elements are significantly similar in concentration. Image by Philip J. A. Skipper.

course, it is possible to determine that precisely through using statistics, which was discussed in Chapter 4.

These graphs offer a flat, 2D representation of data. Sometimes, it is necessary to plot three variables in a more 3-dimensional way. This introduces the concept of the z axis, coming straight out of the page and at a right angle to both the x axis and the y axis, in addition to the standard x and y axes. A good example of this might be presenting differences pre- and post-cleaning in L*a*b* colour space (Figure 5.9). If we think back to the cleaning example given for ANOVA in Chapter 4, an additional useful measure would have been colour change, since plotting the colour changes on a 3D graph would give the reader an intuitive feel for how effective the different cleaning methods were.

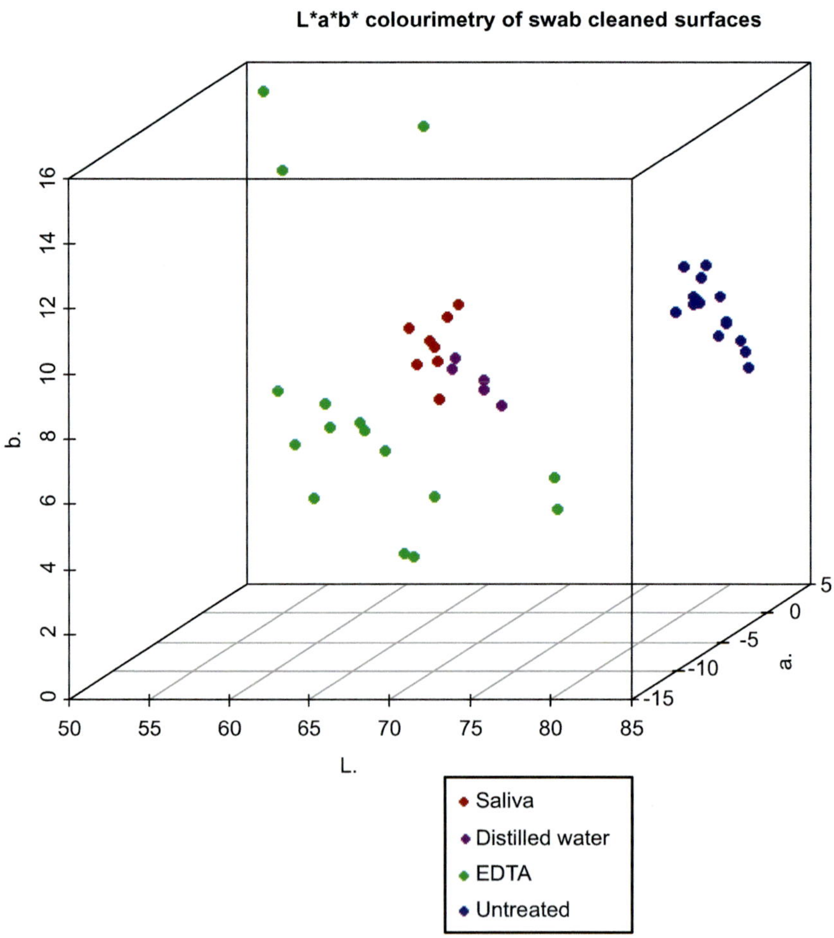

Figure 5.9 Plotting colour in L*a*b* colour space showing a 3-dimensional plot of data. Image by Philip J. A. Skipper.

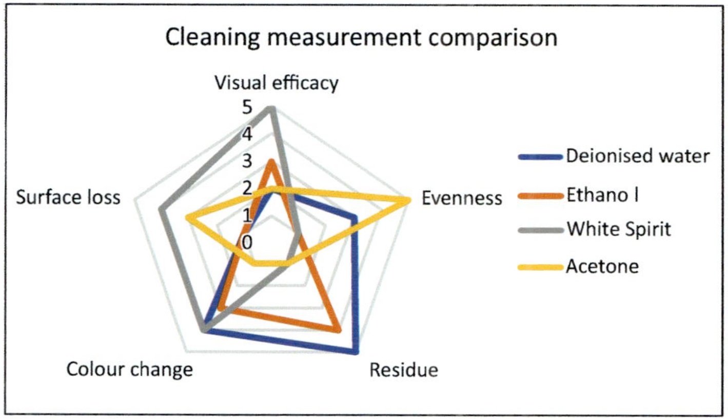

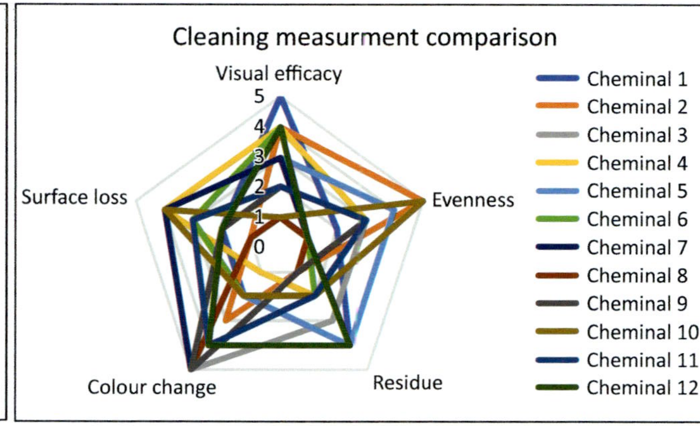

Figure 5.10 Left, radar chart showing effectiveness of four cleaning methods. Right, radar chart showing effectiveness of 12 cleaning methods. This chart is much more difficult to read due to the quantity of information. Image by Lynda and Philip J. A. Skipper.

Radar charts, also known as star diagrams, can be a useful diagrammatic method of presenting several qualities which may be measured as part of a study but are not directly related, although it is important to recognise that they are better for presenting a limited selection of data. If we compare the results of four cleaning methods (this is ordinal data; the data is ranked with no indication of the spacing between the ranks) in a hypothetical study to a radar chart showing 12 pieces of data, we can see that the chart becomes overwhelmed and confusing when too many results are presented in this fashion (Figure 5.10). Radar charts can also be a good method of presenting seasonal or monthly data to show how they compare. Figure 5.11 shows average monthly RH readings from two galleries; the seasonal effects on RH in gallery 2 stand out compared to gallery 1 when presented in this fashion (this is ratio data because it has a true zero).

Henderson *et al.* (2017) is a helpful resource for a range of other graph types, which might be suitable for presenting data in a range of different formats. Although that work focusses on visual depiction of pest-management data, some graphs they mention can also apply to presentation of scientific data.

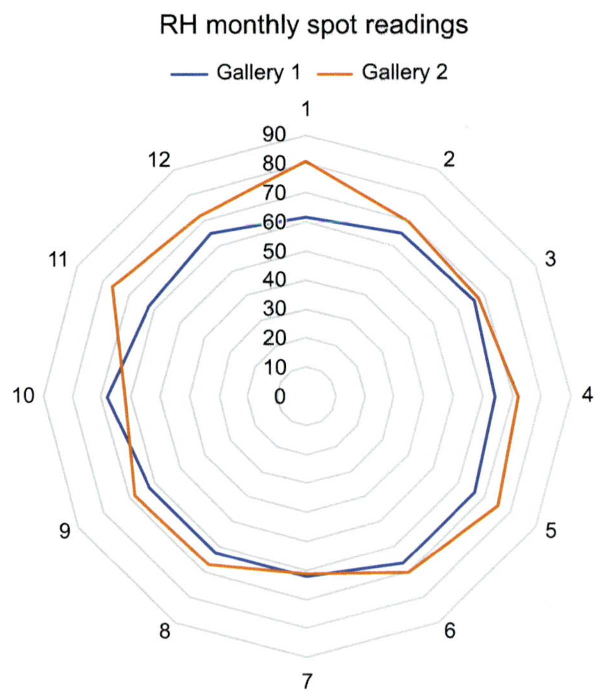

Figure 5.11 Monthly average RH in two gallery spaces shown in a radar chart over a 12-month period. Month 1 is January, 2 is February and so on. Image by Philip J. A. Skipper.

5.5 Presenting qualitative data

Although qualitative data cannot be presented graphically, it is often very suitable for presenting visually. A clear photograph, microphotograph, or drawing provides good opportunities to visually demonstrate an aspect of scientific results.

These visual data may need to have **labels** added onto the figure itself or image processing to add false colours to highlight specific features. A **caption** alone will not necessarily point the reader towards the crucial part of the image. Where any significant post-processing or alteration of the image has taken place, it's important to state this in the methods section of the work so that the way in which it has been changed can be understood. Figure 5.12 shows an example of image before and after post-processing to introduce false colour.

Labels and/or annotations on the image should be short, concise, and proportionate so that they are not too large compared to the size of the image. This means they will not be overly distracting from the image itself. Labels can be in addition to false colour or used as an alternative way of directing the viewer towards a point of interest. The figure should still have a caption as well, which we will talk about in section 5.6.

It is not always necessary to have a single image either. A series of images combined into one figure, similar to that shown in Figure 5.12, may also be more informative than showing one image at a time. Examples may include changes over time or through combining graphs and visual information from one sample.

5.6 Creating meaningful image captions

Captions are the text that explains the content of the image, and they are essential for how people understand any figures presented within the results. The term **figure** is used broadly here to represent all graphs, tables, equations, and images – in other words, any information that accompanies the main text of the report. Scientific journals style all graphs, images, and diagrams as 'figures' but number tables and equations separately. In some humanities subjects such as art history, it is common to name coloured images 'plates' and all black-and-white images, diagrams, and drawings 'figures'. Although we are using the term 'caption' here, it is sometimes referred to as a **figure legend**, or simply legend. As 'legend' is also a term relating to a key within or on a graph or chart, we are using 'caption' to avoid any potential confusion.

A good caption is written clearly but succinctly. Captions should contain enough information to be read on their own, without the reader needing to look at the accompanying text. A handy tip: starting a caption with 'this graph shows' or 'this image shows' is generally redundant, because it's not necessary to tell the reader that this is a graph or image showing something – it should be visually very clear already that this is, indeed, an image. To keep it succinct, that phrase can be removed. Conventionally, the first part or sentence of the caption is the equivalent to a title of that image.

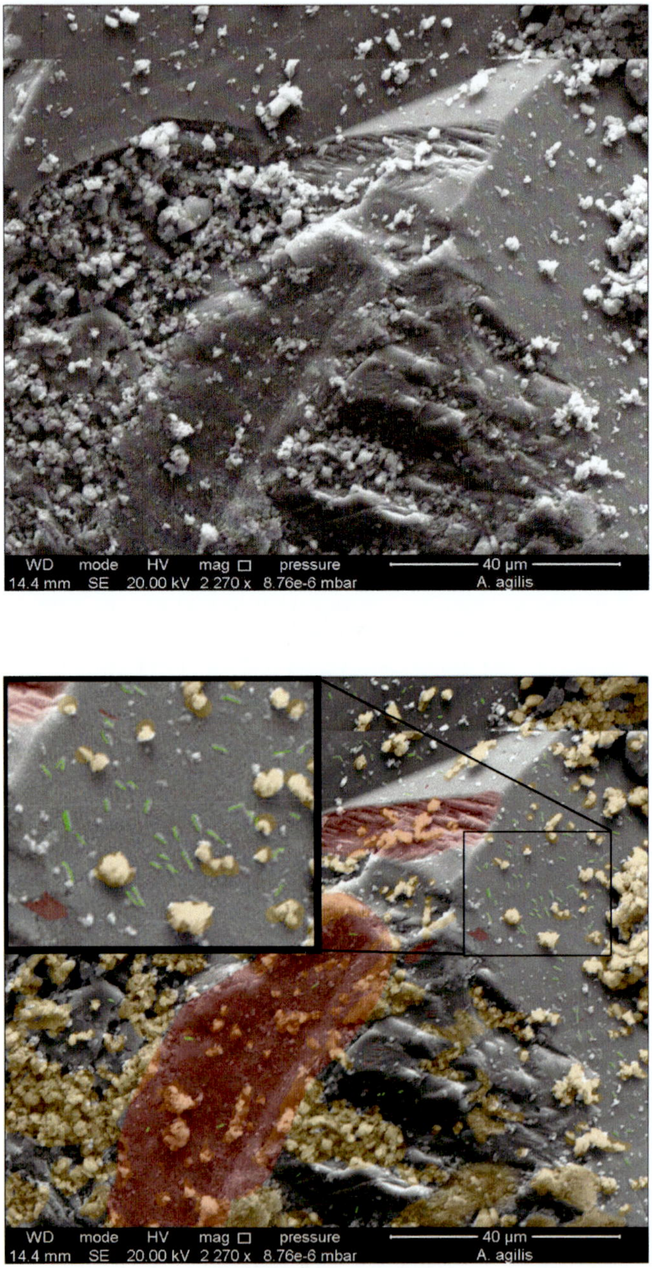

Figure 5.12 Left, Bacterial colonisation of stone, unprocessed image from SEM, magnification x2270. Right, Processed data to add false colour to damage (red), biofilm (yellow), and bacteria (green) and with zoomed-in area to make the green-coloured bacteria more obvious. Image by Philip J. A. Skipper.

There are two potential approaches to writing captions. One approach is when the captions are designed to take on a descriptive format. This means the caption tells the viewer what they are looking at, but it is described in an objective way. The significance of that data is then interpreted further and discussed in the main body of the text. An alternative approach uses the caption itself to draw out some information from the data. This could be viewed as potentially biased, as the writer is inviting the reader to engage with the data in a particular way. However, it can also be helpful, because it links the interpretation of the figure into the overall aims and hypothesis for the work. An example of the same data using both these approaches is given in box 5.1.

As before, the intended audience should be considered in planning **caption** style and format. In the case of writing for an academic journal, the publication may provide its own guidance about caption style, and obviously, this should be followed rather than any personal preferences. As a general rule, in academic writing, the caption should not repeat anything that is stated in the text. Many journals favour short captions.

Box 5.1 Examples of different caption styles

[Caption style 1.] FTIR spectra showing absorbance readings from unknown pigment (red) and lead white pigment (blue).

[Caption style 2]. Comparing FTIR spectra from the unknown pigment (red) and lead white (blue) showed a significant match between the two samples.

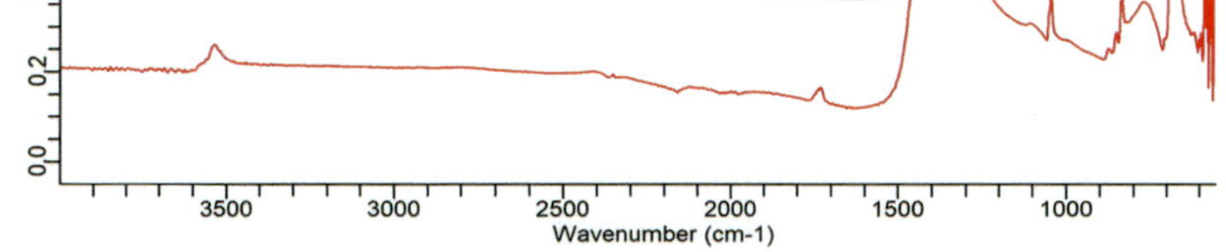

Figure 5.13 Image by Lynda K. Skipper.

The second caption gives detail of how it's interpreted, whereas the first one broadly just states the facts of what is being shown. Note that both give details about which colour relates to which coloured line! Which type of caption do you prefer and why? (There's no right or wrong answer to this question, but it is useful to consider this as your writing style develops.)

Captions may also draw attention to specific details in the figure. In the example shown in box 5.1, the caption could be improved through stating the wavenumbers of the key absorbance peaks so that the reader knows what features are important in comparing the samples. There would usually be a maximum of four short sentences in the caption; these are not designed to be mini essays in themselves! If a microscope was used, it is standard practice to include any magnification in the caption if the image itself does not have a scale bar. In a published paper, where the size selected for the printed image is unknown to the author, it can make sense to state in the caption 'Imaged at x250', for example.

The other captions in this book have been written with this advice for good practice in mind. An interesting personal exercise would be to look through other chapters of the book and see how the captions we have used meet this standard. You may also have ideas for how you would choose to alter or improve them – writing style is a personal preference and everyone has a unique 'voice' or style.

5.7 Summary

This chapter has looked at how data can be presented in engaging ways to enable more effective communication of ideas and information. Having feedback from a colleague can be quite helpful here, as sometimes, a way that seems clear to one person may leave another feeling a bit confused, so do consider asking someone to look over your data presentation plans if possible.

Thinking about how the data might be presented is useful as part of the planning stage to help inform your data collection. However, it's also not unusual for people to collect all the data first then trial a range of different ways of representing that data visually to work out what looks the most effective.

The key point is that data should be shown in a way that aids the overall argument that you are making in the output, so the presentation methods that you choose need to be relevant to support the text. Well-presented data can determine whether your work is viewed as credible or not.

The other part which is important for credibility is, of course, the text of the article or other output. In the next chapter, we'll talk about how this can be structured, considering the desired audience, and what writing a scientific report means. We will also talk more about how to write in an academic style.

Exercises

5.1 A conservator has found the following values for elemental composition of a Duralumin aircraft part through XRF and has drawn a graph of the data. What's wrong with this way of plotting it, and what could they have done instead? Use the data in the table to plot your own choice of graph instead.

Element	%
Aluminium	93.3
Copper	4
Magnesium	1.3
Manganese	0.5
Titanium	0.1
Zinc	0.2
Silicon	0.3
Iron	0.3
	100

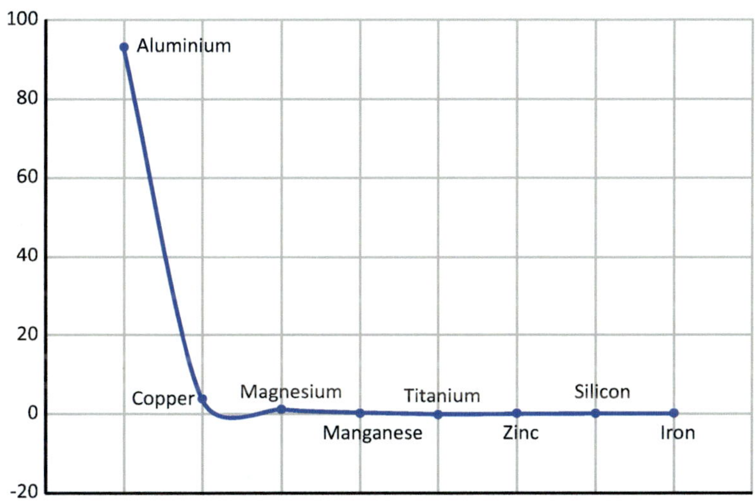

Figure 5.14 Image by Philip J. A. Skipper.

5.2a Look at the latest image that you took on your phone or another device that's nearby, and write down a caption that tells a viewer something about that image.

5.2b Reflect on how you wrote it; did you use an objective style or a more interpretative one? Is there anything you might change now that you are looking at it again?

6 Writing reports in scientific style

Introduction: the importance of dissemination

Carrying out scientific research is potentially useful for your own personal understanding and developing your own practice and methods of conservation. Without **dissemination**, though, that work will not benefit the wider conservation community or enhance understanding and practice more widely, and when you leave the profession, all your hard work in gathering data could be lost. Dissemination is the act of communicating the research that you have carried out to all the audiences that might benefit from it. Some funders of research demand dissemination of the results as a condition of accepting the funding and require proof of dissemination within a specified time after the project itself has ended.

This is where all the work carried out in the book so far needs to come together, to form the **output** of the study, and to make sure that your work is communicated to others. When we use the word 'output', it refers to the production of something which disseminates your findings to others. It's not unusual for some studies to have multiple outputs, with different styles because of different formats. Perhaps there might be a conference poster, a social media post, a longer technical report on your organisation's web page, and an article in the published conference proceedings, all from the same piece of work.

Communication is an important skill. There are a range of different theories about how communication works. One approach is to consider that you are transmitting a message to a receiver. The message is your work, and the receiver is the person who is accessing that work. Communication can be impacted by noise. That might come from the receiver being distracted; perhaps they are thinking about what to have for dinner, perhaps their chair is uncomfortable, or maybe it is just that they do not find the message interesting enough.[1] There can also be noise from the way in which the message has been sent, for example if the style of writing is too complex or unclear. Noise relates to how the message becomes changed or altered so that the receiver does not receive the message that you intended to send. The message – in this case, the information about the findings of your work – needs to be clear and appealing for the chosen audience.

Thinking carefully about how your outputs are planned and then written will improve how successfully your information is communicated. This chapter

DOI: 10.4324/9781003397137-9

considers a range of different standardised approaches that may apply to your planned outputs, focusing mainly on a more formal, academic-style format.

6.1 What is academic writing style?

The phrase **academic writing** is commonly used, but not everyone will necessarily have the same understanding of what the term means. This is because while there are certain typical conventions, it is also something that is generally fluid and flexible. As mentioned in the last chapter, everyone has a unique voice when writing, and we cannot expect everyone's style to be identical. That level of consistency would probably become quite boring when reading through a lot of work! A common mistake, however, is the idea that academic writing needs to be complicated and use as many long words and as much technical jargon as possible. In reality, it is possible to write in an academic style and still be clear and concise through targeting your work to your intended audience.

Standard expectations for good academic writing involve writing in an impersonal style and keeping the use of pronouns – whether 'I', 'my', or 'we' – to an absolute minimum. Box 6.1 gives an example of how the academic writing style can be applied in practice. In this book, we have deliberately chosen to use a less formal style and to address you, the reader, more directly.

Box 6.1 Academic writing style

Writing using personal pronouns, in an informal way:

'Firstly, I added 20 ml of water to the glass beaker, then I heated it up gently on a hotplate until it reached about 30° C. After this, I added five drops of a detergent-and-enzyme solution.'

This style of writing would be absolutely fine in a blog article, perhaps for a public audience. Rewriting these sentences, we can turn it into an impersonal style that does not use 'I', which is typically the style used in academic writing.

'20 ml water was gently heated in a glass beaker on a hotplate until it reached 30° C. Five drops of a detergent-and-enzyme solution were then added.'

You will also note that in box 6.1, the examples were both written in past tense, because it described things that have already happened. That is the standard approach to use with academic writing, though on occasion, it may be necessary to talk about future work or future implications of the findings, in which case changing to a present or future tense is seen as acceptable. The style of sentence structure used in the second example is known as **passive voice** (box 6.2). Writing in a passive voice tends to suggest a more objective, unbiased approach to what is happening because it removes the personal aspect (linking it to the scientific method, which also aims to be unbiased and objective).

Box 6.2 Passive and active voice

In active voice, which is the style people would tend to use when speaking, the subject of the sentence might perform an action (verb); this action is received by something (object)[2] (*Figure 6.1, top*).

Passive voice involves using a sentence structure in which the subject of that action comes first (Figure 6.1, lower). The subject (that's 'I' in this case) has been taken out of the passive-voice sentence. If you are writing this in a method, the reader would assume that you or a colleague heated the water, but that's not relevant – the crucial information for reproducing the method is that the water needs to be heated.

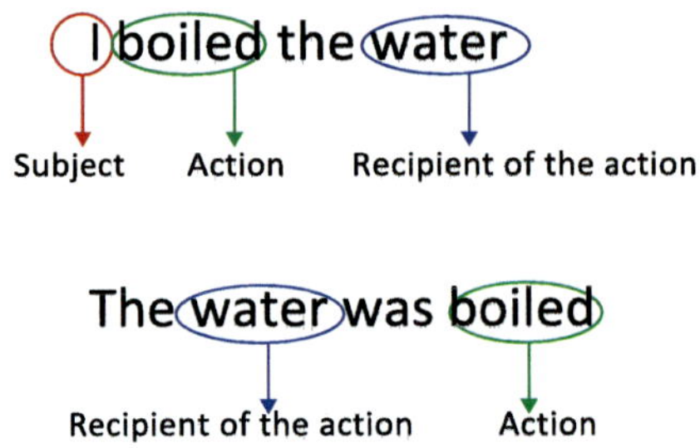

Figure 6.1 Active voice (top) and passive voice (lower) showing sentence construction. Image by Lynda K. Skipper.

The next point we need to consider is that academic writing should be written in a way that is clear and to the point. This was mentioned in Chapter 3, when we discussed how being precise can help to build confidence in the validity and reliability of the work. We also talked about this in relation to writing good captions in Chapter 5. The first sentence in this paragraph was deliberately *not* very concise; 'Academic writing should be clear and concise' says exactly the same thing. The main point of the sentence is still clear, but the second version uses fewer words.

Academic writing tends to use formal language in other ways, not just pronouns and sentence structure. This means avoiding contracted words. For example, when writing, you would use 'could not' instead of 'can't', and 'is not' instead of 'isn't'.

Colloquial or locally used expressions may not be understood by a potentially international audience. In a nutshell, don't write your sentences like this, okay?[3]

Abbreviations are always linked to the full text the first time they are used, for example 'scanning electron microscope (SEM)', and then the abbreviation alone should be used afterwards, including in the captions. The exception to this is particularly common abbreviations in the field, such as RH for relative humidity, which may not need to be written in full for a conservation audience. Being aware of your audience will help you to decide on a suitable approach; Philip co-authored an article about the effects of RH on COVID in heritage interiors for the *Historic England* research magazine, and the editor decided to explain RH because it was aimed at the general public (Skipper and Willett, 2021).

What is absolutely key is that the output builds an argument. Building an argument involves presenting and evaluating the data from your research in a logical order to build evidence that moves towards a conclusion – the equivalent of the end of our book or the end of a novel. Sometimes, an argument will evolve during the course of writing the work, at which point it can be edited to make the argument more explicit through all the sections. At other times, the argument will become clear as data from the research is analysed and the writer knows what they wish to persuade the reader of from the beginning. Having a strong argument will help to shape all the sections of the output, and it can be useful to separately note down ideas about that argument as soon as they start to emerge.

The flow of text is a concept which is quite difficult to describe in a book. It's about how a piece of work moves from one topic to the next and takes the reader along with it. When written work has good flow, these transitions are managed in such a way that one section clearly leads into another, and it's possible to see how the argument is developing over time. When transitions of topic are not managed in that way, the work can feel disjointed or unconnected. To help with this, each section would often have a sentence or two that links into the previous section, and each end of the section should likewise have a sentence that leads into the next one. Generally, one concept or idea would be presented per paragraph. Again, this helps with conceptualising and structuring ideas in a way that aids the flow of the work by not over-complicating or repeating information.

Academic writing is in itself a skill. Not everyone is good at academic writing straight away, and it certainly does not come easily to everyone. You wouldn't expect to be able to pick up drum sticks and do a perfect drum solo the first time; it would take a lot of practice to be able to get to that point. With academic writing, good writing is a constant process of learning. As part of the accreditation process for conservators in the UK and some other countries, conservators demonstrate their commitment to learning through continuing professional development (CPD); learning in conservation does not stop after graduating from a degree programme. Writing ability is also a skill which can be improved through learning and practice. In part, this learning is self-led; as you write, your writing tends to improve simply through the act of writing itself. Feedback from others, such as colleagues, tutors, or through peer review of submitted articles, is an important part of that process of continuous improvement. Reading other people's work and considering what you like and styles that don't work as effectively for you will also help with understanding different writing styles and formats. Reading information about writing in books or other publications, such as this one, will also help with the process – but it is certainly a skill which is best developed by 'learning through doing'.

6.2 Audience and outputs

Different audiences require different approaches to communication, and different writing styles form part of that, as well as thinking about how data is presented, as discussed in Chapter 5. It's therefore a useful skill to be able to adapt your style to the target audience.

In Table 6.1, there are some suggestions of how your choice of writing style might be adjusted to fit different audiences. This is not a comprehensive table, so you can add to this yourself. What is your planned audience, and how might you adjust your writing style to fit that audience's needs?

Table 6.1 Types of outputs shown with their potential audience and some considerations for communication styles.

Venue	*Type of audience*	*Style*
Conservation conference presentation	Conservators and conservation students	Professional but engaging and less formal. Can use professional terms without much explanation, though technical knowledge of specialist subjects may be more limited.
Blog on museum website	Mostly general public, adults	Informal, explaining technical terms when necessary – but try to keep them minimal. Usually only shows selected data, not all of the research findings.
Specialist journal for that subject area	Specialists in that field, potentially other researchers in related fields	Academic, formal language, using technical terminology with often minimal explanation.
Internal museum report	Professionals in the museum who may not have any specialist knowledge of your area	Usually formal language, explaining terminology. May focus more specifically on key points/recommendations given first, not at the end as a conclusion.
Social media post	Broad – public and professional audience, mainly adults	Generally very short, focusing on one or two key points, often with visual material. Tends to use less formal language.
Text panel in museum display	Broad – public audience, may be adults and children	Short with a strict word count, quite informal, often focused around key points. Assumes very little prior knowledge. May have links, e.g. to museum website for more detail.

As a reader, papers or reports may seem complex to read if you are not the intended audience; for example, reading a paper outside your area of specialism in order to understand how a technique has been approached in a different discipline or when you are reading about something that is new or unfamiliar. If the paper was written with one specialism in mind, it may take several readings through the paper to follow it fully and/or reading around the topic to help you understand the paper better. That can still be a useful approach to reading articles even when it is a subject you feel more familiar with. When we review papers, we usually read through multiple times – with an initial skim-read to pick up the main points, or argument, of the paper, then at least twice more, reading it closely to see how they build and evidence their argument to ensure we have a very thorough understanding of what the author(s) is discussing.

6.3 Traditional structure of a scientific report

In this section, we use the term 'report' to mean any output from your work which needs to be written in an academic style, which may include a dissertation or thesis and articles for publication as well as a report either for internal use or for publication.

The way in which a report is structured follows an academic writing format. In writing stories, you will generally come across the idea that they should have a beginning, a middle, and an end. Scientific writing is no different; this, too, should have a clear beginning, middle, and end. The way it differs from a novel and an essay is in how the sections are laid out. In reports, there are clear and informative headings, setting out expectations for what will be in each part. Longer reports and books may need a contents page and a list of figures and/or list of graphs and tables. Articles do not usually have the indexing system with the lists of contents and figures, but they will still follow a set of headings.

A typical format could be:

- Abstract
- Introduction
- Method
- Results
- Discussion (not always necessary)
- Conclusion
- Notes (not always necessary)
- Acknowledgements
- References
- Appendix or supplemental data (if required)

Obviously, there is some flexibility here based on the content and topic being discussed. There may also be some variation depending on the instructions from journal editors and publishers. For example, although a literature review may be included in the introduction (which is how it's discussed in the section that follows), sometimes, it is more appropriate for it to be a separate section or written as a separate chapter, particularly if it is part of a longer piece of work such as a dissertation or thesis.

Results are not necessarily one section; often, there may be multiple subheadings or section headings that fall under results. These are being considered here as a block of information, since again, this layout will vary depending on the type of output being produced.

The first step of writing is to prepare a structure, using headings and sub-headings as needed. The different sections are not usually written in order from beginning to end, as a planned structure means that sections can be populated without needing to start at the beginning. It's not unusual for the introduction to be a fairly late addition, and the abstract, which is listed first, is actually written last (because it provides a summary of what is in the report). One of the hardest parts of writing can be at the very beginning, and box 6.3 contains some tips for getting started!

In the next part of the chapter, we will talk about what these different sections are composed of.

Box 6.3 Motivation, or how to start writing

Sitting down and looking at a blank white screen can be daunting. Students often ask us how they can start, sometimes because it's difficult to decide when to stop reading and when to start writing and sometimes because they don't know the best place to start or even where to begin. A few tips that we have found useful are:

- *Make headings and subheadings first.*
- *If the word count is really tight, give a moment's thought to how long each section ought to be. This may prevent you writing such a detailed introduction that there are no words left for a vital section such as the conclusion!*
- *Start by writing the section that you feel most comfortable with.*
- *Start with completing your methods. You know what you did, and it just needs describing. This is one of the simplest sections to write if you have recorded your steps thoroughly.*
- *Make a GANTT chart to plan how you will use the time you have available, and refer to it regularly to make sure you stay on track.*
- *Set yourself mini targets for each day or writing session.*
- *Reward yourself when you reach milestones, whether that's a daily target or a weekly one. Decide what reward would be best at motivating you.*
- *If you have something you're not sure of, don't interrupt the flow of writing by going to look it up. Make a comment in the margin or a coloured highlight, and come back to it later.*

- *Don't expect work to be right the first time. Getting something written down even if it's not quite right gives you a basis for editing later.*
- *Don't expect work to be perfect, ever! Work that is 'the best you can do at the time' is a more realistic aim.*
- *Take a break when you need one.*
- *Sometimes, things don't work as planned. Be kind to yourself if you aren't meeting your planned timing, reflect on the reasons, and ask for help and support if you need it.*

6.4 Abstract

The abstract is often the last part of the report to write, because it is a summary of the whole document. In order to be able to write this, you need a very clear idea of exactly what you are saying throughout and what the key points are from the work. Abstracts are generally fairly brief (200–300 words for articles; one page, which is 400–500 words for a thesis), and contain what we would describe as the 'spoilers'. If this were a murder mystery novel, the **abstract** would be the equivalent to an initial section just before the start of the novel saying who the killer was and what happens with the plot twist at the end of the book.

Even though this is often the last thing you write, it is usually the first thing your audience will read. A poorly written abstract that does not clearly communicate the contents of the article will discourage someone from reading it. Equally, an abstract can be a useful filtering mechanism when someone is looking to read papers on a specific topic. If the abstract shows that it's not going to be relevant, then the reader can move on to look at other papers. It helps to save time and makes the process of reviewing relevant literature more efficient.

Some conferences require a **paper proposal** so that the presenters can be selected by a technical committee, and after the conference has happened, presenters write and submit papers. Here, the abstract usually differs from the paper proposal: it will be shorter, and it should summarise the results obtained. The paper proposal might instead have outlined what was going to be investigated – and clearly, it had no results to report but instead defined a research question or a hypothesis.

6.5 Introduction to the report: including aims and objectives

An **introduction** is designed to give an overview of the field and to talk about the purpose of your research. In the case of the book, we have used introductions at the start of each chapter alongside an overall introduction at the beginning. For articles, you would just have a single introduction.

This is where your hard work in refining aims and objectives is also helpful; the introduction is based in part around this aim or aims and what you are setting out to prove through the work. To show that the aim is worthwhile, the introduction needs to set out something that is referred to as the context. This is the wider field in which your research question sits. You need to use this wider review of literature in the field (a literature review; see 6.5.1) to demonstrate that your research is novel and that you are aware of how others have contributed to the field. It may also be appropriate to discuss the methodology that you have used in your research within the introduction.

An introduction should therefore aim to set out the purpose of the research and the key aim(s) and objectives and discuss why it is a necessary project that contributes to overall knowledge in the field. You can think of it as your marketing section – you are trying to sell your 'product' i.e. the article to a peer reviewer, so you need to convince them that your article is worthwhile by being clear about how and why that work makes a valid contribution.

An **introduction** may also be used to set out the **scope**. This is where you define clear boundaries for your research. These boundaries may be geographical or person specific, for example looking only at ormolu produced in France or only furniture produced by Alvar Aalto (1898–1976). In those cases, ormolu produced in Germany or furniture produced by Charles and Ray Eames would be out of scope. Alternatively, boundaries may relate to a particular time period, such as the 17th century, or era, e.g. Ancient Egyptian. It may also be quantity related, such as examining 10 objects from a collection of 500. Think carefully about how you phrase scope in relation to time constraints. Even if you could only look at ten objects because you only had one week available, you can enhance that justification through showing that the sample is representative of the wider collection.

6.5.1 Literature review

Demonstrating your understanding of the wider field of study is done through a **literature review**, which was also mentioned in Chapter 1. At that point, you were encouraged to use literature to develop your research proposal and research questions. For the report, you need to write about what you discovered, and you may wish to also read more widely to help with that. A literature review should be approached with the same standards of scientific rigour as your experimental work, and include accurate references to show whose work has informed your own.

A literature review aims to establish the gap(s) in current knowledge. It demonstrates that your work is necessary, because it covers something that has perhaps not been done before or not been applied in that way to that field of study. It can be used to show how your own research helps to advance understanding of the topic. It also helps to build confidence in your work by showing you have a thorough understanding of other work in the field.

There are some key aspects to a literature review. It should be a **critical analysis** that summarises, compares, and evaluates information. A critical analysis does not mean that you need to criticise others; you need to make judgements, but they need not be judgements that everyone else's work is terrible. In making a critical analysis; perhaps author X has looked at subject Aa, author Y has looked at subject Ac, and when viewed together, it identifies there is a gap in knowledge in subject

Ab, which is where your own work fits into the overall picture of research in the subject.

Because it is a summary, it would be unusual for a literature review to discuss one author per paragraph. Literature reviews aim to draw together information from multiple sources to allow them to be compared. That comparison can then lead to an evaluation, which brings it together in a way that supports your overall argument. In other words, what is missing or not being said? Sometimes, you might identify issues with repeatability or in the methodology of others, which may mean conclusions drawn by those studies are not correct. Again, this would be an area for discussion within the literature review.[4] An example of how this might work in practice is shown in box 6.4, which compares two literature sources on RH guidance for museums.

Box 6.4

Arts UK, a charity which provides online access to the UK's art collections, recommends that RH should be controlled within a range of 40% to 65% RH, with fluctuations of no more than +/-10% within a day (no date). They do not list specific materials. However, the Canadian Conservation Institute guidelines give different RH bands for different object types, such as keeping most metals at 45% to 50% RH (CCI, 2021). This conflicting advice highlights the potential difficulties in understanding the correct environmental parameters for storage and display.

Green = summary text
Red = comparison (sometimes you don't need much to make it a comparison)
Blue = evaluation

We tend to subdivide literature into primary sources and secondary sources. **Primary sources** are the direct evidence of a topic, such as objects, documents written by someone who was present at an event, or papers based on original datasets. **Secondary sources** are the processed (analysed) information, such as journal articles which give a review of a topic or discuss a series of historic documents to give a perspective after the event. It is useful to consider the source of the data and any potential areas of **bias** or inaccuracy that might introduce. A contemporary account of an event that someone did not agree with will have a different perspective and bias to someone who felt positively about that same event.

Another division is that between scholarly, peer-reviewed literature and **grey literature**. This includes materials such as government documents, working papers, reports, blogs, and company information. In general, grey literature is not peer reviewed and can vary hugely in terms of quality and content. Therefore, you should be cautious when using it and check for potential bias and inaccuracies. Grey literature can also have a limited period of availability, so if you are planning on referring to a report or blog post, it is important to save a personal copy! Sources which compile information (such as Wikipedia and Science Direct's topics page) are valuable tools for identifying sources, but you should always follow through to the original papers that they quote and check information yourself. Inaccuracies could have been introduced by the person or people summarising the information from the original piece of work. You should always reference the original source or a peer-reviewed review paper instead of Wikipedia. With peer-reviewed literature, you can be more confident in its accuracy, it is less likely to be biased, and the publishers have a financial incentive to ensure that it remains available.

6.6 Methods section

The **methods** section is very factual, describing 'how' things are done. This is often the section that people find easiest to start with, since it can be written alongside the experimental work. This part is also very important for ensuring others can repeat the work you have completed, using the same methods, and obtain the same result. If the method is not completely clear, perhaps omitting one important instruction by mistake, it could prevent that person's work from being replicated.

Methods do not always need to be detailed in full. When you are following an established method, you can simply cite the original publication that gives that method in full. If we were to carry out an Oddy test of a material,[5] we would not need to write out the full details of that Oddy test, just cite the protocol used. It wouldn't be enough information simply to state that it was tested using an Oddy test, though, as there are a range of variations on the original method from 1973. Citing the exact source means that the specific variation of the Oddy test can be clearly linked to the method that the researcher used. If everything is following a standard method, and the key aspect of the report is not the method but the findings, then it is not always useful to have a long methods section. In journal articles, the method section is still usually expected to be present, however short, but for some other types of output, it may be possible to combine the method and result sections.

In the methods section the authors usually record details of equipment used, and sometimes the materials and suppliers as well, although some journals ask for a separate list of suppliers at the end. You will need to know details for everything, such as which company supplied your ethanol. Found in the back of a cupboard in a jar labelled 'ethanol' by the technician or a colleague may be an accurate description of the origins of the ethanol that you used, but it generally wouldn't be appropriate to share that in an article. Hopefully, you would be able to look back on ordering information to work out exactly where it came from! Recording this in a lab book at the time will make this section *much* easier to write. Trying to work out supplier details for chemicals and which microscope (and magnification range and lighting conditions if not standard ones) you used perhaps six months or a year after you did the research makes it more challenging to be accurate.

6.7 Results section, or, how not to just write down all the information from the graphs again, but in words

This section has quite a long title but one that we felt was quite relevant for the section in question! **Results** are the information obtained from your study; experimental data, charts, tables, images. Presenting these was discussed in Chapter 5. When talking about the results, the information in the written text-based section needs to complement the information that is presented visually. Crucially, it should not involve simply writing down the same things again. The text-based section of the results should pull out key points and emphasize significant findings. In the case of FTIR spectra, for example, stating where the key peaks are and what chemical bonds those might relate to links the visual results (where the peaks should be labelled) to the interpretation.

Results need to clearly link to the main argument and the purpose of the testing. Presenting results without any context for why that particular test has been done can leave readers confused about how it fits with the aim of the work.

As with the first chapters of this book, we tend to think of the division between results and conclusion as something that can also be approached as 'what?' and 'why?'. The results section covers 'what?'; talking about what has been observed from the experimentation. There may be some limited assumptions or interpretation taking place at times, such as relating FTIR peaks to chemical bonds and perhaps stating some type of meaning – perhaps a previously unidentified component is a protein-based adhesive, for example. 'Why?' is discussed within the conclusion (see section 6.8) and covers the significance of that result and how it links into the overall story or argument that is being made through the paper. Box 6.5 gives an example of how this might be applied.

Box 6.5

We can use the method of considering which piece of information is a 'what?' and which is a 'why?' to help us to decide where that information might go in the report. Look at the following paragraph, and consider which parts are a result and which are conclusions.

Figure 6.2 *shows a magnified image of cracking and loss to the glaze of a porous ceramic. This damage may be due to temperature and humidity changes, including moisture penetrating the glaze, or be due to a manufacturing fault.*

Result (what): Figure 6.2 shows a magnified image of cracking and loss to the glaze of a porous ceramic. This is a 'what' because it is describing what is visibly present.

Conclusion (why): This damage may be due to temperature and humidity changes, including moisture penetrating the glaze, or a manufacturing fault.

This is a 'why?' because it gives some explanation for why the damage might have happened.

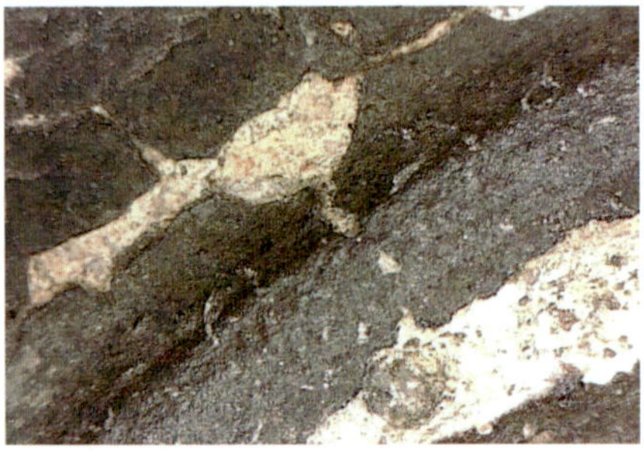

Figure 6.2 Crazing and loss of a brown glaze on an earthenware ceramic, magnification x230. Image by Lynda K. Skipper.

Although it is normal for there to be some limited interpretation of results in this section, it can sometimes be challenging to decide how far to take the interpretation of the results and their meaning in terms of the overall aim of the study. This may vary depending on the type of output, and any instructions for authors can give specifics relating to particular guidelines or approaches. It is important not to repeat yourself across both the conclusion and the results sections, so keeping the theme of 'what' in mind is generally a good starting point for how to approach results.

6.8 Conclusion and/or discussion section

You may already have noticed that some journals use **conclusion** and **discussion** as separate headings, and some use just one of those definitions. Both of these terms relate to the 'why?' discussed in the previous section. Where discussion and conclusion are present, the expectation is that the discussion relates specifically to the discussion of the results, and the author should interpret those findings and show their meaning at this point. The discussion is a useful place to discuss uncertainties. Where there are potential areas that might be open to more than one interpretation for example, they need to be acknowledged here. Any limitations of your methodology, such as limits of detection of instrumental analysis, might be relevant. The conclusion can then be used to draw together inferences about the wider context or explanation of that information and how it can be used to gain broader understanding of the topic. The difference can be quite subtle, and it's certainly something that we have found challenging to write in the past when deciding what part of a discussion/conclusion to place where. A summary of how these sections could be divided is presented in box 6.6.

Box 6.6

Discussion

- *Interpret results and their meaning*
- *Potential uncertainties in understanding*
- *Potential limitations of methodology*
- *Summarise overall argument*

Conclusion

- *Relate to overall aims and objectives of the research*
- *Place results in a wider context*
- *What additional research might be required*

Where only one of those headings is used, all those aspects are discussed under the same section. In either approach, it's important to relate them back to what was covered in the introduction and to again revisit that core argument that runs through the paper. The way that your results have been presented and structured should lead the reader through your thought process to allow them to (hopefully) reach the same conclusions as you did. By revisiting the project aims and objectives in your summary section, you can discuss whether the research project has met those initial plans, as set out in the introduction.

A conclusion or discussion is not the place to introduce new results; you should only be talking about what has been presented in the previous sections, with one key exception. In any study, there may be side projects or other tests that could have been run, but there was perhaps not the time, funding, or both to include them at that point. Perhaps there are still some identified gaps in knowledge that could be filled by further work in that area. This section gives the writer the opportunity to highlight those, as prompts to encourage others to continue the work, or to perhaps assist their own future funding bids. However, the conclusion should emphasise what has been discovered rather than what has not. Stating 'More research is needed' is not a conclusion: the reviewer and later the reader will instantly think 'Why didn't that author go off and do more research, then?'

6.9 References

Use of relevant references is crucial for showing your understanding of other work in the wider context of your field of interest. References are also useful for the reader to expand their understanding through reading related information in their field of interest.

There are two different terms that can be used here, references and bibliography, which are sometimes used interchangeably. However, this is not technically correct, as **references** are works which you have consulted and then cited in your writing, whereas a **bibliography** is for sources which have shaped or informed your work in some way, but you have not directly cited them in the text.

References are used to acknowledge sources and work completed by others and should accompany every point or idea which is not your own work. Careless or poor referencing can mean that someone reading your work is unable to find the information that's been mentioned, so they cannot follow and gain understanding of that topic further. Failing to reference work entirely is **plagiarism**, the theft of someone else's ideas. Using appropriate references helps to build your own credibility as an author and creates confidence in your work. It's not always appropriate or necessary in every format; this book is not designed to be written as an academic article with detailed referencing of every point. Therefore, in this book, you will see occasional in-text references, which we have combined with other sources (our bibliography) as a section for further reading at the end of the book. Another term that is used is **citation**, which is the short cue or pointer in the text to the full reference.

There are a variety of different referencing styles, and in our experience of publishing articles, every conservation journal seems to want a different referencing format. Therefore, we are not going to talk about particular referencing styles here, but what is crucial is that the style is consistent and accurate. It saves a lot of time if the full reference is captured when first obtained, including details such as the editors of a book as well as the authors of the chapter that was relevant to your research, the year of publication, and the name of a publisher such as 'IIC' in full. Automatic referencing software can be a great time-saver, but it should not be fully trusted to produce an accurate reference list every time, because sometimes, information does not go into the right boxes automatically. We have found several instances in which the reference is all in capital letters and we have to correct it manually!

6.10 Acknowledgements

Naturally, **acknowledgements** are important to say thank-you to anyone who has helped. An informal approach to acknowledgements is more common in dissertations and theses (perhaps thanking family, friends, and even the local pizza delivery service), while published papers would generally take a more formal approach to acknowledging those who were professionally involved in the published work. It is important to acknowledge support given for analysis, perhaps through training or access to equipment, or throughout sourced work. Contributions from other stakeholders, such as object owners or representatives from communities of origin for objects should be noted in this section. The writers should also acknowledge any funders. When reading papers, it's useful to check this section and to consider whether the funding bodies could lead to any potential sources of bias. A famous (or perhaps infamous) example of this is from the 1950s tobacco industry, as discussed by Brandt (2012), where the papers funded by the tobacco industry showed a much more positive viewpoint of smoking.

6.11 Appendices and supplemental data

Appendices are more commonly found in assessed work, such as a thesis, or in a book. These usually contain additional information which is not covered in the main body of work, such as raw data or additional images. In publications, we usually find the term **supplemental data** used to refer to this type of information. It may be uploaded to a website or data repository. Including raw or supplemental data is generally a positive approach where that data exists. It allows greater transparency of findings and means that work can be re-analysed in future, perhaps in light of new knowledge. It may also reduce the risk of data loss. If information is stored on only one computer hard drive, for example, malware or hardware failure may make that data inaccessible. Some funders of large projects demand that raw data be made available through publications.

If using some kind of external hosting, it is always sensible to check who can access the data and who retains copyright and ownership of the information. There is a risk with any type of electronic data storage that file formats may become obsolete over time. If information is stored in a file format linked to a particular manufacturer's software, which can only be opened with that same brand of software, it may no longer be supported by the latest operating systems. It may also cause problems with accessing data if other researchers do not have a copy of that particular software. The preservation of electronic data is a conservation field of its own, and detailed discussion of all the aspects involved in this are outside the scope of this book.

Considering how to present this data in a way that is accessible for future use may mean spending time typing up numbers written in a lab notebook, converting file types into a different format, and making sure that tables of data and charts contain meaningful headings that explain the context. A large sheet of numbers is not meaningful unless enough information is given about the source and type of data.

6.12 Keywords

For many submissions, the authors are asked to choose **keywords** that are relevant to their work. Generally, five or six keywords are used. These are used as metadata to link into search engines and allow people searching for specific topics to find the output. In order to help others find your work, it is important to get these right. Potential keywords range from very general and generic to specific around topics and themes or techniques. Some potential options for these are shown in Table 6.2, though this is not a comprehensive list of all possible keywords. Keywords need not be a single word, as can be seen in the table, but they should not be a long sentence either.

Table 6.2 Keyword ideas divided into a range of different themes.

Theme	*Potential keywords*
General conservation/heritage	Heritage science; conservation; heritage; art history; museums; collections management; archaeology . . .
Subject/material specialism	Paintings; books; paper; textiles; built heritage; industrial objects; natural history; preventive conservation . . .
Date/era	5th century BCE, Han dynasty, later 20th century . . .
Location specific	Scandinavian peat bogs, Tibetan plateau . . .
Topic specific	Climate change; sustainability; artists' materials; biodeterioration; nanomaterials . . .
Technique specific	FTIR; SEM; spectroscopy; XRF . . .

When considering suitable keywords, it is best to ask yourself the question: 'What would I look for if I was trying to find work relating to this topic?' You may even find inspiration from the words you used initially in carrying out the literature review! One generic keyword is quite useful, because that links your work into the broader field, and the remainder would usually be based around the more specialist terms specific to your research relating to era, topics, materials, and techniques, as suggested in Table 6.2. Brainstorming words may help with this, although it's not unusual to find that this approach generates a lot more keywords than are allowed for the submission, and some filtering will be needed to reach the required number!

6.13 AI in scientific writing

We are seeing an increasing proliferation of **artificial intelligence (AI)** tools alongside increasing levels of debate over their future in a range of different fields and what is considered ethical use of these tools within writing. Many universities and journals will have their own policies around how AI can be used, and these should be consulted before considering its use in producing outputs. For people who are writing in a non-native language or who perhaps find spelling and grammar difficult, AI may be able to help through providing some editing of sections of a manuscript or assisting with making a sentence or paragraph sound more academic in style. It is also possible to input a sentence and ask AI to rephrase it to make it clearer or to summarise it to reduce the word count. In general, while spelling and grammar checking (such as carried out on this manuscript using Microsoft Word) would not be acknowledged in a manuscript, more comprehensive assistance with writing and phrasing would generally need an acknowledgement, whether that is done by a colleague or by AI. Even so, AI is not perfect and should always be re-checked for inaccuracies, and the writing style used by AI may not fit well with the author's own style. This impacts the flow of the text through the change in writing style, making it clear in the text where an AI contribution has been made and detracting from the quality of the work overall.

Using AI to generate data or other results is, of course, at the other end of the spectrum, as it is obviously unethical to use technology to falsify or invent results. Using AI-generated text directly, without paraphrasing or attributing a source, is also plagiarism, a topic also discussed in the referencing section. However, there is still a middle ground where AI has potential uses, but the ethics of that approach is less clear. Moving beyond assistance with what is generally described as copy editing, AI has the potential to write sections about different concepts or ideas when given a specific prompt or carry out a review of literature on a topic. Current issues found by our own testing of AI indicated that it was often repetitive when asked about a topic and did not go into much detail. It often repeated the initial prompt, but in a slightly different way. Some references used by AI may be incorrect or may not even exist at all, as found by Altmäe and colleagues (2023). At present (mid-2024), using AI to write sections of a manuscript based on a prompt risks introducing inaccuracies, which may impact the confidence of the reader (and the reviewers and the editor) in the whole document. Where AI is used, the contribution of AI must be attributed, just as any author would expect to have their work acknowledged.

6.14 Summary

Carrying out brilliant scientific experimentation in conservation is important, and carrying out suitably rigorous work is necessary to a professional conservator, but without good dissemination of the results, no-one will be able to appreciate or learn from that work. Having effective communication skills in a range of different mediums is a valuable asset. Being able to talk about and share research may present barriers due to cost and time restrictions, and people may be nervous of their work being publicly 'judged' through a peer-review format or at a conference.

We hope that this book will help to give individuals the confidence to carry out research, write scientifically, and publish their work by having a process to follow to design and carry out experiments relevant to their work. Breaking down a science-based output into sections was designed to show that scientific writing is something which follows a set structure, and through following the structure, it is possible for someone with limited scientific experience to produce a science-based and evidence-based output. In many ways, the writing process mirrors the scientific method itself, as it is a circular process of development and self-reflection alongside feedback, which allows the final output to be shaped and refined, ready for its publication.

Exercises

6.1 Transform the active voice sentences below into passive voice. Could the wording of any of them be changed to a more academic-sounding style?

  a Our results show a clear difference in colour of the surface before and after cleaning.

b We added 50ml distilled water to the sample, and then we heated it on a hotplate for 20 minutes.

c After heating, I used a piece of pH paper (1–14 scale) to check the pH of the sample.

6.2 A conservator has carried out research to study the mosaic inlay materials used for a Pre-Columbian Andean mirror. A number of different shell types were used, with some stone. What keywords would you use in a publication about the findings?

6.3 Separate the elements of this extract of writing into the method, results, and conclusion.

A Motic BA10Pol polarising microscope was used, and images were taken using the manufacturer's software. This showed that the pigment was a yellow lake, which is an organic dye precipitated onto a transparent alum base. The images showed there were some of the alum-based pigment particles present, along with a white filler which appeared to be chalk (calcium carbonate).

Notes

1 This is not necessarily a reflection on your work; if someone is not interested in that topic, they are unlikely to engage with your message, no matter how good your communication style is – you can't expect to appeal to everyone.

2 'Object' in this case is being used as a grammatical term for something that receives an action; it is not an object in a heritage or conservation sense.

3 'In summary, do not write a sentence like this' would be a more academic-style alternative.

4 Try to be nice and respectful about it, though; 'this author's work has some limitations' is a kinder way of phrasing it than 'this author's study was rubbish'.

5 Oddy tests are used to find out whether a material produces gases that could cause increased deterioration to objects, often those made from metals.

References and further reading

AIC. (1994) *Our code of ethics*. Available at: www.culturalheritage.org/about-conservation/code-of-ethics.

Allende, J. (2004) 'Rigor – the essence of scientific work', *Electronic Journal of Biotechnology*, 7(1). Available at: www.ejbiotechnology.info/index.php/ejbiotechnology/article/view/1112/1494.

Altmäe, S., Sola-Leyva, A. and Salumets, A. (2023) 'Artificial intelligence in scientific writing: a friend or a foe?" *Reproductive BioMedicine Online*, 47(1), pp. 3–9.

Andrews, R.J. (2019) *Info we trust: how to inspire the world with data*. Hoboken, NJ: Wiley.

Brandt, A.M. (2012) 'Inventing conflicts of interest: a history of tobacco industry tactics', *American Journal of Public Health*, 102(1), pp. 63–71.

CCI. (2021) *Caring for metal objects*. Available at: https://www.canada.ca/en/conservation-institute/services/preventive-conservation/guidelines-collections/metal-objects.html

Chan, E., Aambø, M., Łojewski, T., Grimstad, I., Ufnalewska-Godzimirska, M. and Sandu, I.C.A. (2022) 'Microfade lightfastness testing on *The Scream* versions in the Munch Museum collection', *International Journal of Conservation Science*, 13, pp. 1623–1634.

Clark, K. (2001) *Informed conservation: understanding historic buildings and their landscapes for conservation*. London: English Heritage.

Feller, R. (1994) *Accelerated aging*. Los Angeles: Getty Conservation Institute.

Freedman, J.L. and Sears, D.O. (1965) 'Selective exposure', in Berkowitz, L. (ed.) *Advances in experimental social psychology* (vol. 2). New York: Academic Press, pp. 57–97.

Hair, J., Black, W.C., Babin, B.J. and Anderson, R.E. (2010) *Multivariate data analysis*. 7th edn. Upper Saddle River, NJ: Pearson Educational International.

Henderson, J., Baars, C. and Hopkins, S. (2017) 'Trends in effective communication of integrated pest management data', in Bridgland, J. (ed.) *ICOM-CC 18th triennial conference preprints, Copenhagen, 4–8 September 2017*. Paris: International Council of Museums, pp. 137–155. Available at: www.icom-cc-publications-online.org/1780/Trends-in-effective-communication-of-integrated-pest-management-data.

Icon. (2019) *Ethical sampling guidance*. Available at: www.icon.org.uk/groups-and-networks/heritage-science/heritage-science-resources.html.

Icon. (2020) *Ethical guidance*. Available at: www.icon.org.uk/resources/resources-for-conservation-professionals/standards-and-ethics/icon-ethical-guidance.html.

Ireland, C., Skipper, L. and Elie, M. (2014) 'Cleaning of metals contaminated with historic pesticide residues that are hazardous to health', *Journal of the Institute of Conservation*, 37(1), pp. 65–81.

Luxford, N. and Thickett, D. (2011) 'Designing accelerated ageing experiments to study silk deterioration in historic houses', *Journal of the Institute of Conservation*, 34(1), pp. 115–127.

McMillan, K. and Weyers, J. (2011) *How to write dissertations and project reports*. 2nd edn. London: Pearson Education.

NASEM. (2019) *Reproducibility and replicability in science: 5. Replicability*. Washington, DC: National Academies of Sciences, Engineering, and Medicine etc., National Academies Press. Available at: www.ncbi.nlm.nih.gov/books/NBK547524/.

Noss, R. (2007) 'Values are a good thing in conservation biology', *Conservation Biology*, 21(1), pp. 18–20.

Odegaard, N., Carroll, S. and Zimmt, W.S. (2005) *Material characterization tests for objects of art and archaeology*. London: Archetype Publications.

Reedy, T.J. and Reedy, C.L. (1992) *Principles of experimental design for art conservation research*. Los Angeles: Getty Publications. Available at: www.getty.edu/conservation/publications_resources/pdf_publications/principles_experimental_design.html.

Salkind, N.J. (2007) *Statistics for people who (think they) hate statistics*. New Delhi, India: Sage.

Sinek, S. (2011) *Start with why: How great leaders inspire everyone to take action*. London: Penguin.

Skipper, L. and Rubinstein, N.A. (2018) 'Detecting chloride contamination of objects and buildings: evaluating a new testing process', *Journal of Conservation and Museum Studies*, 16(1), pp. 1–9.

Skipper, P.J.A. and Willett, C. (2021) 'Effects of relative humidity on Covid-19 in heritage interiors: historic England', *Historic England Research Magazine*, May. Available at: https://historicengland.org.uk/whats-new/research/effects-of-relative-humidity-on-covid-19-in-heritage-interiors/.

Stuart, B. (2008) *Analytical techniques in materials conservation*. Chichester: Wiley and Sons.

Walliman, N. (2019) *Your research project: designing, planning, and getting started*. 4th edn. Washington, DC: Sage.

Wisker, G. (2018) *The undergraduate research handbook*. 2nd edn. Basingstoke: Palgrave Macmillan.

Wright, D.B. (2002) *First steps in statistics*. London: Sage.

Answers to exercises

1.1 This will vary because different researchers may take a different approach. One answer could be:

Aim: Evaluate methods for identification of woods in Shinto statuary.

Objectives: Review different methods used in literature.

Use a case study approach to test a selection of methods on the statues.

Evaluate their effectiveness (accuracy, cost, speed, ethical implications) and make recommendations for most suitable techniques to use.

1.2 Hypothesis – Auxilium Conservare is a suitable replacement for DCM.

Null hypothesis – Auxilium Conservare is not a suitable replacement for DCM.

1.3 We used Google Scholar and found that 'Indian coin analysis', 'archaeological coin India', and 'composition coins of ancient India' were good starting points for generating a range of background information and different techniques used in analysis.

Box 1.3 answers

Hypothesis: The intensity of the green pigment used on Victorian green wallpapers directly correlates with the levels of arsenic present.

Null hypothesis: The intensity of the green pigment used on Victorian green wallpapers does not relate to the arsenic concentration.

This one is less clear-cut because the researcher as a person is not changing a variable, simply measuring arsenic and colour on an undefined number of green areas of wallpapers. This is still a valid hypothesis and null hypothesis and is a good example for when having an independent variable is not required.

Hypothesis: Increasing ultraviolet light levels causes more rapid fading of textiles.

Null hypothesis: Increasing ultraviolet light levels does not cause more rapid fading of textiles.

We can assume here that the textile remains a constant, and it is the ultraviolet light level which changes. This means the textile is the dependent variable, and the UV light is an independent variable. To make it more precise, the researcher could specify what type of textile and dye.

Hypothesis: Addition of zinc oxide nanoparticles to silicate-based stone consolidants reduces biofouling.

Null hypothesis: The presence of zinc oxide nanoparticles in silicate-based stone consolidants has no impact on biofouling.

In this case, the zinc oxide nanoparticle levels would be changing. It also implies more than one type of silicate-based stone consolidant. Both would be independent variables, and the biological growth level would be the dependent variable.

2.1 We are not giving an answer to this exercise, because it is based around your own personal experience. After you have completed it, take some time to reflect on what you've proposed, and re-visit the relevant sections in this book. How might you improve or refine your plans further?

2.2 same as 2.1.

2.3a Literature review, elemental analysis such as XRF.

2.3b Literature review, multispectral imaging, FTIR, or GC-MS.

2.3c Literature review, and FTIR is commonly used for plastic identification.

2.3d A spot test can be used to identify lead, and that would answer the research question given. If doing a test specific for lead, it would not detect if any other potential health hazards are also present which perhaps the person asking the question had not considered. XRF would often be used to check for a wider spectrum of hazardous chemical elements.

3.1 Bone alone would show whether the bone itself without filler showed a colour change. If possible, the filler could also be tested when not in contact with the bone through making up some filler samples in a small inert plastic or glass container. Three repeats of each sample would be needed to check for changes to the control.

3.2 The student has a few different options available, such as looking for historic pewter that is sacrificial. This is likely to be the closest match to the original material, though again, it raises questions around ethics. Modern pewter could be bought, but the chemical composition of modern pewter is different due to the absence of lead. Carrying out tests on just tin would have a similar limitation around how transferrable results are to historic pewter. In this case, using repeated testing on modern pewter might be helpful to optimise a smaller number of methods before testing on a sacrificial piece of historical pewter. Carrying out multiple tests with repeats on the modern samples reduces the testing required on a historic piece, as one repeat could then validate the technique(s) as being appropriate, or not, on the historic item.

4.1 Mean: Add up all the weight data and divide by 5, which is the number of weights listed.

(15.6 + 20.02 + 18.9 + 110.5 + 19.1) / 5 = 36.8 kg (1 decimal place)

Median: Line up all the numbers in order, and choose the number in the middle. 15.6, 18.9, 19.1, 20.02, 110.5. The central one is 19.1 kg.

These numbers are quite different, and how useful they are would depend on the question being asked. Working out an average weight per object to design shelving would mean that using median alone, it could lack the capacity to hold the heavier objects. However, if the research was to look at typical characteristics for objects donated in 2023, the median (or mode) may be a more useful statistic. In this case, as there is only one of each weight type, mode cannot be used simply from the data as it is presented, but categorising objects into weight groups, e.g. 0–10kg, 10–20kg, would give three items in the 10–20kg category, and therefore, that would be the mode value.

4.2a Qualitative
4.2b Quantitative
4.2c Qualitative and Quantitative
4.2d Quantitative
4.2e Qualitative
4.3a Nominal
4.3b Ordinal
4.3c Interval
4.3d Ratio
4.4a Chi-square test
4.4b Chi-square test
4.4c ANOVA
4.4d Student's *t*-test

5.1 The student has drawn a line graph. These data points should not be connected to each other because they are not related; therefore, each one needs to be shown independently. As the data gives a percentage composition, a pie chart could be appropriate here. Alternatively, a bar chart could be used, though the proportions of the smaller elements means the bar sizes would not be clear. It may be necessary to have a two-part scale, omitting a section in the middle. The benefit of a bar chart is they could also show error bars from the XRF reading, thus giving a clearer idea of the confidence of those results. This would also be an example of when to use standard deviation for error bars because you are not trying to compare the aluminium with the zinc etc. If you also picked up on the lack of y axis labelling and no caption on the graph, well done!

5.2 This exercise aims to help you develop your own style, and there are no incorrect answers here.

6.1a The results showed a clear difference in colour of the surface before and after cleaning. (This one didn't need rearranging, just to have the pronoun removed. We've also adjusted it to the past tense. 'There was a clear difference in surface colour after cleaning' is a more concise way of saying it.)

6.1b Fifty millilitres of distilled water was added to the sample before heating it on a hotplate for 20 minutes. [Note that as the first word of the sentence is now a number, generally, we write out the number as a word unless this would be very lengthy.]

6.1c When heating was completed, pH paper (1–14 scale) was used to measure the pH of the sample. (In this one, 'measure' has been used instead of 'check', as it is more formal phrasing.)

Your answer may have slight variations, and that is fine – just take the time to reflect on how you answered and whether it meets the academic writing guidance we talked about in the chapter.

6.2 Object-specific (mirror) and time/location-specific (Pre-Columbian and/or Andean), material-specific (shell/stone), technique-specific (mosaic). The conservator could also use keywords giving the ways in which they were analysed, though the specific techniques are not listed in this example.

6.3 'A Motic BA10Pol polarising microscope was used, and images taken using the manufacturer's software' is the method.

Results would be the image itself (which is not shown here) along with any tests carried out to help identify the pigment. The identification of pigment, alum base, and chalk would be a result.

Conclusion would discuss what the lake pigment is, i.e. manufacturing technique, and reason(s) for why the chalk is seen (as a filler).

Glossary of terms

Abstract summary of the entire contents of the output
Academic writing formal writing style generally using passive voice
Accelerated ageing *see* 'artificial ageing'
Acknowledgement place to thank anyone who has had some input or influence but is not a co-author of the output
Active voice sentence construction in which the subject of the sentence is doing the action, e.g. I ate a cake
Aim(s) the outcome you hope to achieve through your work
Analysis of variance *see* 'ANOVA'
ANOVA a hypothesis test to determine whether the differences in groups of data are significant
Appendix, appendices additional or supplemental information that is not included in the main text
Argument(s) the main idea from the findings of your research, which you would then support with evidence
Arrhenius equation equation to calculate reaction rates with changing temperature, used in artificial ageing
Artificial ageing subjecting a material to high temperatures, RH, light levels, or other conditions to replicate many years of exposure in a short space of time
Artificial intelligence (AI) a computer-based system which performs a particular task simulating human intelligence
Bar chart(s) graph showing different categories as rectangular bars, in proportion to their values
Bias inclination towards a particular outcome of the research, which may lead to contradictory results being overlooked
Bibliography list of sources which have been consulted and informed the work but are not cited in the main text of the output
Caption information detailing the contents of a figure, graph, or table, placed either directly underneath or above the item. Sometimes known as figure legend
Case individual units which make up the sample
Chi-square a hypothesis test for qualitative data
Citation in-text reference to a source of information
CoBRA (conservation-based research and analysis) the research, analysis, survey, and investigation necessary to understand the significance and inform decisions about the aspect being studied
Compound light microscope an optical microscope that uses more than one lens to enlarge the item being viewed
Confirmation bias inclination towards a particular outcome which reinforces previous beliefs or assumptions

Conclusion(s) Summary of the findings of a study drawing together information to present a final argument
Control(s) aspect that remains unchanged during the testing, used to see whether the variable has caused an effect
Critical analysis evaluating evidence to reach a judgement about your own and others' work
Data processing the act of carrying out classification or analysis of information
Dependent variable an aspect that changed which is being measured during the testing
Discussion summary of the findings of a study drawing together information to present a final argument; if a conclusion is also present, the discussion would summarise the area studied, and conclusion would place it in a wider context
Dissemination the process of sharing information with a wider audience
Documentation use of recording techniques to capture information
Electromagnetic spectrum The full wavelength range (or frequency range) for electromagnetic radiation
Electron negatively charged sub-atomic particle found within atoms
Element(s) a substance that cannot be further broken down to smaller parts by chemical means, such as silicon, carbon
Error bars a line representing the amount of uncertainty or variability in a value
Ethical sampling transparent, accountable decision following an ethical framework
False positive a test result which is incorrectly showing as positive
Figure(s) any information that is presented visually to accompany the main text of the report
Figure legend *see* 'caption'
Fourier transform infrared spectroscopy (FTIR) analytical technique that uses infra-red light to detect bond vibrations as part of identifying components
Gas chromatography – mass spectrometry (GC-MS) analytical technique that separates chemical components in a sample that has often been dissolved in a reagent or vaporised and detects their mass/charge ratio as part of identification of each chemical compound
Glass transition temperature (T_g) the temperature at which a substance changes from a solid state to a soft or tacky state, sometimes a range of temperatures
Grey literature literature that is not peer reviewed, such as blogs, reports, personal websites, or some (older) conference proceedings
Hypothesis (plural: hypotheses) a prediction of an outcome, which is then tested by research
Independent variable the variable that is altered by the researcher to test the effects
Infrared (IR) area of the electromagnetic spectrum with a slightly longer wavelength than visible light
Inorganic Items that are not classified as organic, i.e. they do not contain both carbon and hydrogen, and their chemical behaviour is not due to the presence of those elements
Interval data that can be categorised with a rank, and it is possible to infer equal intervals between neighbouring cases, but there is no true zero point
Keyword(s) words or short phrases which are used as search terms for a topic
Kurtosis a descriptive measure of the sharpness of the peak of a data series
Labels description of a visual aspect on a figure or graph to aid understanding
Line graph graph where individual data points are connected by a line
Literature review(s) evaluation of the published research available in your field of study
Mean the average value, calculated by adding all the numbers together then dividing by the total amount of numbers
Median the mid-point of a list of numbers sorted into numerical order
Method(s) a way of carrying out a process of research

Methodology the reason or justification for using a particular method to carry out the research
Microfading accelerated ageing of a very small area to infer light sensitivity; uses a high-intensity light with a spectrophotometer
Microscope an instrument used for creating enlarged images of an area
Mode the most common number or category of data
Multispectral imaging (MSI) equipment that collects a series of images at different light wavelengths from across the electromagnetic spectrum
Negative control(s) test that is expected to give a negative result; allows for checking of method
Nominal labelled groups of sample data with no particular order
Normal distribution symmetrical distribution of data around the mean (average) with most data close to the mean; looks like a bell curve when plotted on a graph
Null hypothesis a hypothesis stating that any differences in the testing results are not caused by the variable that is being tested
Objective(s) measurable steps within a project that are carried out in order to reach the aim
Observation gathering information by looking at something
One-tail a hypothesis test that looks at whether the sample mean is higher or lower than the population mean but does not test both simultaneously
One-way ANOVA a hypothesis test that explores whether the mean of one independent variable is different from the population mean
Open-ended distribution class of information without a boundary, e.g. less than 5 or more than 2,000
Ordinal ranked categories of data without information as to the interval between the categories
Organic items containing carbon and hydrogen chemical elements, where the main chemical behaviour of the substance is based on the presence of carbon
Outlier(s) a data point that appears different from the main body of data
Output(s) the end product of the research, generally used to communicate the findings
Passive voice sentence construction in which the subject of the sentence is acted on by the verb, e.g. the cake was eaten
Photogrammetry use of photographic images to generate a 2D or 3D model
Pie chart(s) a circular graph divided into sections showing the relative size of that 'slice'
Pilot study a small initial study, with results used to inform the process of the main research project
Polarised light a filtered light wave that only vibrates in one direction
Polarising light microscope (PLM) a microscope with filters known as the polariser (in the lighting path) and the analyser (inserted in the light path from the sample) to create polarised light; used to characterise optical properties of the sample
Population(s) the whole group of subjects or objects that we are interested in
Positive control test that is expected to give a positive result; allows for checking of method
Primary source(s) an original account of the topic, which includes analysis of raw data
Processed data information which has been analysed in some way (see also data processing)
***p*-value(s)** a statistical measure of whether the likelihood of an event happening is simply due to chance
Qualitative non-numerical data such as colour, location
Quantitative numerical data such as height, surface roughness
Radar chart a visual depiction of multiple numerical variables on a single scale which all radiate from a shared central starting point
Raman spectroscopy analytical technique which measures light scatter due to rotation and vibration of bonds

Ratio numerical data with a true zero point
Raw data data before any processing has been carried out
Reference(s) a source which has been cited in the text of your work
Referencing the act of citing a source
Reflectance transformation imaging (RTI) a photographic method to generate surface images with an interactive, variable light source
Repeats the act of carrying out testing more than once; a minimum of three repeats is conventional in scientific testing
Reproducible something that can be copied or repeated
Research process leading to the discovery of new knowledge
Research paradigm values and beliefs that form the basis of your approach to a research project
Research question the problem that your research is intending to address, not necessarily phrased as a question
Results data from a study or research project; may be used as a report heading
Rigour the process of designing testing in a way that is likely to give accurate results
Sample(s) a smaller group selected from a population in statistics or a small part removed for experimental testing
Scanning electron microscope (SEM) microscope which uses electrons instead of light to create the image; can reach very high magnifications
Scanning electron microscope with energy dispersive X-ray spectroscopy (SEM-EDX) SEM with additional analytical tool that uses X-rays to determine the elements present in a sample
Scientific method using experimentation to test a theory in an objective way
Scope the area or topic that the research covers
Secondary sources information which was created after the event; may review existing work in the field
Significant probability of the differences in the data being due to chance; typically, a *p*-value less than 0.05 is statistically significant
Skew a measurement to describe whether the distribution of data is symmetrical or asymmetrical around a mean
Spectrophotometer a device to measure colour
Spot test(s)(ing) a chemical test to determine whether a particular element or compound is present in a sample
Stakeholder individuals or groups who have an interest in the end result of the research
Standard deviation a measurement of the amount of variation within a single group of data
Standard error (SE) a measurement of the variation in a population, derived from the standard deviation
Star diagram *see* 'Radar **chart'**
Stacked bar chart a graph in which each of the bars are subdivided into different sections
Statistically significant *see* 'significant'
Stereo microscope binocular optical microscope, generally with a low magnification range, typically x10–x100, and typically with a good working distance
Student's *t*-test A hypothesis test for statistical significance between two groups of data
Surrogate material(s) modern or disposable materials which are used as an alternative to original heritage materials, with the same properties as the original
Surrogate object an object which is used as an alternative to the original heritage object but has the same properties as the original
***t*-test** *see* 'Student's *t*-test'
Trend line a line showing the general tendency of data points on a graph
Tukey's honestly significant difference *post-hoc* test (Tukey's HSD) A test to identify which of the groups is significantly different from the mean; used alongside ANOVA

Two-tail a hypothesis test that looks at whether the sample mean is higher or lower than the population mean

Two-way ANOVA a hypothesis test that explores whether the means of two independent variables are different from the population mean

Ultraviolet (UV) Short-wavelength light which forms part of the electromagnetic spectrum

Valid, validity how well a study accurately measures the aspect that was being tested

Value(s) (1) the meanings or qualities assigned to heritage or approach to scientific study; (2) a number or set of numerical data points

X-ray diffraction (XRD) analytical technique that uses X-rays to measure the crystal structure of a substance

X-ray fluorescence (XRF) analytical technique that uses X-rays to determine the elemental composition of a substance

X-ray Short-wavelength light which forms part of the electromagnetic spectrum; used in some analytical tools or alone to view inside objects

Index

For Product Safety Concerns and Information please contact our EU representative GPSR@taylorandfrancis.com Taylor & Francis Verlag GmbH, Kaufingerstraße 24, 80331 München, Germany

Batch number: 10722108

Printed by Printforce, the Netherlands